"Linda Civitello takes her readers on an interesting and learned journey about a little-known subject: The history of leavening agents. I'm grateful for this detailed backstory on what makes bread rise to the occasion."

—Adrian Miller, author of *Soul Food: The Surprising Story of an American Cuisine, One Plate at a Time*

"It's just an innocuous white powder in a can at the back of the closet. Or is it? Linda Civitello's history shows how baking powder precipitated vicious competition in big business, raised concerns about chemical adulterants in food, and transformed home cooking. Without baking powder, there would be no fluffy cakes and pancakes, no biscuits, no muffins, and no cookies; in short, no American cuisine as we know it."

—Rachel Laudan, author of *Cuisine and Empire: Cooking in World History*

"*Baking Powder Wars* is an insightful and fascinating account of the advent and subsequent struggle for legitimacy of one of today's most widely used ingredients in both home and commercial baking. Linda Civitello succeeds in making what might have been an academically dry topic come alive with erudition, grace, and humor."

—Nick Malgieri, author of *Bake!*

"Linda Civitello has mined her subject thoroughly and documented how it changed American baking to satisfy our hurry-up attitude toward life and food in general, as we embraced a quick and easy solution to the tiresome problem of supplying the family table. Along the way, Civitello records in detail the fundamental history of a business that is almost uniquely American—the baking powder business. Who knew that baking powder could be such a rich resource?"

—Nancy Harmon Jenkins, author of *Virgin Territory: Exploring the World of Olive Oil*

"Baking powder was not a topic high on my interest level before I read this book, but I'm fascinated by the subject now. A great job of getting into the topic and placing it in a broader perspective."

—Andrew F. Smith, editor of *The Oxford Encyclopedia of Food and Drink in America*

Baking Powder Wars

HEARTLAND FOODWAYS

The Heartland Foodways series seeks to encourage and publish book-length works that define and celebrate Midwestern food traditions and practices. The series is open to foods from seed to plate and to foodways that have found a home in the Midwest.

A list of books in the series appears at the end of this book.

Baking Powder Wars

The Cutthroat Food Fight That Revolutionized Cooking

LINDA CIVITELLO

Manufactured in the United States of America
1 2 3 4 5 C P 5 4 3 2 1
∞ This book is printed on acid-free paper.

Library of Congress Cataloging-in-Publication data
Names: Civitello, Linda, author.
Title: Baking powder wars : the cutthroat food fight that revolutionized cooking / Linda Civitello.
Description: Urbana : University of Illinois Press, [2017] | Series: Heartland foodways | Includes bibliographical references and index. |
Identifiers: LCCN 2016055808 (print) | LCCN 2017014066 (ebook) | ISBN 9780252099632 (e-book) | ISBN 9780252041082 (hardcover : alk. paper) | ISBN 9780252082597 (pbk. : alk. paper)
Subjects: LCSH: Baking powder—United States—History. | Baking powder—Economic aspects—United States—History.
Classification: LCC HD9330.B23 (ebook) | LCC HD9330.B23 U525 2017 (print) | DDC 338.4/766468—dc23
LC record available at https://lccn.loc.gov/2016055808

The story [of the baking powder war] . . . is quite as absorbing in its way as the story of the rise to power of the Standard Oil Company. It shows the working of a mind quite as masterful as that of a Rockefeller and equally unscrupulous. At the same time it is the story of the most gigantic advertising ever attempted in this age of big advertising schemes. Its self-confessed sponsor must be given the title of the greatest advertising genius of the century.

—"Baking Powder and Boodle," *Spice Mill* magazine, October 1903

It isn't alone sobriety, industry, and honesty that make success, but battle, too.

—Lincoln Steffens, "Enemies of the Republic," *McClure's Magazine*, April 1904

[William Ziegler] declared that he would suppress our [baking powder] industry, even if it cost him his life or his fortune.

—William Ziegler obituary, 1905

Contents

Acknowledgments

This book would not have been possible without the kindness of many strangers and good friends at libraries, in academia and publishing, and in the food history world. I am lucky that I live in the country that invented public libraries, and I am deeply indebted to every librarian I have ever met in my life. The custodians of culture without whom this book would not have been possible are at the Rhode Island Historical Society, repository of the Rumford papers; Tammy Popejoy at the American Institute of Baking in Kansas City, Kansas; the student and career librarians at UCLA, and the good people of the state of California for creating it; ditto to the public library in the City of the Angels; Marylee Hagan at the Vigo County, Indiana, Historical Society and Museum; Sean Eisele, Special Collections Librarian at the Vigo County Public Library; Megan Marvin at the Clabber Girl Museum; and the Kraft Heinz Company, which provided clear copies of early twentieth-century Calumet cooking pamphlets and advertising.

On the internet, indispensable is Jan Longone, genesis and curator of Feeding America: The Historic American Cookbook Project at Michigan State University. My FB food colleagues Jonelle Galloway and Elatia Harris, and Toni Tipton-Martin's Jemima Project website were constant sources of information. *Baking Powder Wars* is much richer because the internet has made ephemera available in ways never before possible. I am grateful to the strangers who emptied recipes and cooking pamphlets out of their attics, basements, and kitchen cabinets and drawers and sold them on eBay; to Eric Chaim Kline, Bookseller; to the Antiquarian Book Sellers' Association of America; and to Lulu.

At UCLA I thank Professor Mary Yeager for her high standards and feedback, especially her admonition not to lose track of the women; Assistant Professor Jan Reiff for asking questions that sent me into deeper areas of baking powder research; and Professor Michael Roberts for letting me audit his food law class. At the Harvard "Capitalism in Action" conference, Eli Cook gave me the good news, and Professors Richard John and Thomas Andrews provided valuable feedback.

In the food history world, I am deeply and always grateful to Andy Smith, who is to food history what Meryl Streep is to acting, for his comments and support. The Culinary Historians of Southern California, especially Nancy Zaslavsky and Richard Foss, and Cathy Kaufman of the Culinary Historians of New York gave me the opportunity to present my research to their membership for comments. When the exceedingly magnanimous Nach Waxman found out that I was writing a book about baking powder, he invited me to his bookstore, Kitchen Arts & Letters, and brought out binder after binder containing hundreds of baking powder trade cards. His personal collection of these and thousands more food-related trade cards now resides in the Special Collections Library at his alma mater, Cornell University. I thank Heather Furnas, PhD, Division of Rare and Manuscript Collections, Carl A. Kroch Library, Cornell University. Bruce Kraig—His Bruceness—who became interested in this book while I was his cicerone and chauffeur and he was my captive audience during his Los Angeles hot dog odyssey for *Man Bites Dog*, brought *Baking Powder Wars* to the University of Illinois Press.

At the University of Illinois Press, I thank William Regier and Marika Christofides for their guidance and especially their patience. I am also deeply grateful to my copyeditor, Jill R. Hughes, for catching me when I stumbled; to Roger Kevin Cunningham for his copy; to Heather Gernenz for her marketing acumen; and to Jennifer Holzner for capturing the essence of the book in her cover design. I cannot thank enough my project editor, Tad Ringo, for his phenomenal attention to detail and his miraculous ability to untangle kite strings.

If I had any sense, I would have chosen a topic I could have researched within five miles of my home in Los Angeles. But once the baking powder wars bit me, this Neapolitan was possessed by her own personal tarantella. It drove me to do the research dance in Rhode Island in a January blizzard and a hotel fire, and in Kansas and Indiana during a tinderbox summer of record drought and temperatures. This book is my *Babette's Feast*—one big blowout after years of research.

I could not have done this without my Vassar sisters Barbara Hartley, Pam Ferron, Ellen Hill, and Randi Sunshine; my biological brothers Mike and Joey; my niece Dana and sister-in-law Sue; and my cousin Lenny Trenchard. The antidote for pulling the weekly all-nighters that were the only way I could write this book while working full time was the humor of Mike Hill, Tim Lyons, and Rodney Pinks. Grazie to my paisan, the excruciatingly funny playwright and actress Dina Morrone, for her plays *The Italian in Me* and *Moose on the Loose*. Invaluable were the support of Carol Lynch, Melinda Arnold, Sheridan West, Daniela Galarza, chef Chris Lauderdale, and Kristin Callaghan, Rob Stenson, and their amazing offspring, Aidan and Keira. For their hospitality, I am indebted to Kathy Campbell and Bob Swarthe, Ralph and Kelli Kenol, and Mary Ryan. Nan Kohler of Grist & Toll supplied me with heirloom flour, especially the fragrant Sonora, which I cannot live without now. I also thank chef Adrian Lipscombe for sharing her peach cobbler recipe for this book; and Breadologie pastry shop in Granada Hills, California, where I felt it was my duty to do research. For being a for-real life saver, I thank Dr. Solomon Hamburg; and for repeatedly pulling me out of quicksand and plying me with licorice, usually at the same time, I thank my steadfast, optimistic neighbor, Juanita Lewis.

I am deeply grateful to Barbara Ketchum Wheaton, who lit a torch and blazed the way for the rest of us.

I am sad that Kelly Hill, Jack Arnold, and my cousin and first friend, Sandy Civitello Connors, did not get to celebrate the completion of this book.

But I am glad that my Aunt Yolanda Seneca will get to read *Baking Powder Wars* in her second century.

Baking Powder Wars

Introduction

Business is war. Cooking is chemistry. Food is political.

This is obvious to us in the industrialized twenty-first century. But it wasn't obvious in the preindustrial nineteenth century, when most businesses were small partnerships, chemistry was a hobby, and the largest branch of the United States government delivered the mail. *Baking Powder Wars* delves into what happened before government regulation, when the force controlling the market was not Adam Smith's invisible hand, or Alfred D. Chandler Jr.'s visible hand, but advertising sleight of hand.

The baking powder wars are the Hundred Years' War of business. Baking powder entrepreneurs are the robber barons that history missed. Two of them are directly connected to endowed chairs at Harvard. This Hobbesian struggle for Darwinian survival of the fittest was a series of food business wars: advertising, trade, legislative, scientific, judicial. The baking powder wars were fought among companies with the same formula, among companies with different formulas, between government agencies, in the legislature of almost every state and territory in the United States, in the United States Congress, within federal agencies, and between federal agencies and baking powder businesses. There were wars among female consumers loyal to different brands, in cookbooks, and among women in the new field of home economics. The baking powder wars reached to the United States Supreme Court, to the White House, and into kitchens worldwide.

What Is Baking Powder?

Baking powder is an odorless white powder. But it is a chemical leavening shortcut as revolutionary as the discovery of yeast fermentation that created leavened bread in the Fertile Crescent approximately six thousand years ago.[1] Baking powder is the reason that the expression "flat as a pancake" has no meaning for Americans: it makes the difference between a flat French crêpe and a fluffy American pancake. Feeding a particularly American need for speed, baking powder created new fast baked goods such as baking powder biscuits, cookies, and quick breads like gingerbread, pumpkin bread, and banana bread. Baking powder made cakes, pancakes, waffles, doughnuts, crullers, and muffins easier to prepare, cheaper, and shortened their cooking time radically. It also changed their texture and allowed Americans to increase their consumption of sugar, flour, and fats. Today McDonald's, Kentucky Fried Chicken, Denny's, International House of Pancakes, Dunkin' Donuts, Starbucks, Mrs. Fields Cookies, Sprinkles Cupcakes, and thousands of other multinational businesses would have severely limited menus without baking powder.

Baking powder biscuits are uniquely American: bread and cake, savory and sweet. They are the dumplings in chicken and dumplings, the biscuits in biscuits and gravy, the cake in strawberry shortcake. Dropped on fruit and baked, they are the topping on cobbler. Split open, they make sandwiches and sliders. Baking powder biscuits are as connected to America's national identity as the sports that were invented contemporaneously with them: baseball, basketball, football.

Cuisine or Chemical?

Baking powder is important because its existence poses a basic question: What is food? It is something that nourishes? Or simply something we ingest? Chemical additives in food are routine in the twenty-first century, but they were new in the nineteenth. Baking powder, bleached white flour, and saccharin, invented at approximately the same time, were some of the products that broke down the dam for the chemical flood in foods. *Baking Powder Wars* illustrates how and why food in the United States became not farm to table, but factory to table.

Baking Powder Wars examines four major baking powder companies that arose in the nineteenth-century United States: Rumford, Royal, Calumet, and Clabber Girl. Their founders are representative of nineteenth-century entrepreneurs. They were all white, male, Christian, and educated. They were all self-made men, first in the antebellum sense of men who followed their own visions, and later in the Gilded Age sense of being financially successful. Three of the men had ties to Germany: one was born there, one had parents who were born there, one was educated there. Although baking powder became the destination for all four men, they began their journeys in different professions: academia, grocery, pharmacy, sales. All four baking powder companies began as partnerships between a chemist and a salesman or businessman. In only one case, Rumford, was the chemist the dominant force.

A Note on Food History

Ralph Waldo Emerson wrote: "Man is explicable by nothing less than all his history."[2] So far, that history has been incomplete because it has not included food. Classical philosophers regarded the olfactory organs connected to food—the nose and tongue—as primitive because they were connected to bodily functions. Sight was the privileged sense because it was connected to the higher faculties: thinking, reading, the brain. When people emerged from Plato's cave, they saw the light; they did not smell the bacon. Socrates condemned cooking to the scholarly underworld for two thousand years when he declared it was a craft—*tekne*—not an art. Even in art, still life painting with food was at the bottom of the hierarchy.[3] It is ironic that historians have overlooked food, because food has always been big business. In 1917 only two companies in the United States—United States Steel and Standard Oil of New Jersey—had greater assets than the meatpacking companies Armour and Swift.[4]

Even most anthropologists were not interested in food, because in primitive societies food preparation was manual labor, the work of women. In more advanced societies, cooking was done by women and slaves, often eunuchs. The anthropologists who studied food preparers were men, "who didn't find such matters especially interesting."[5] The feminist historians after the cultural revolution of the 1960s did not want to study topics connected to traditional women's roles.

The semiotics of food have always been available to anyone who would look, but until recently, no one would. As food scholar Anne Romines has said, "Let them read cake."[6] *Baking Powder Wars* reads cake—and cookies, muffins, pancakes, waffles, doughnuts, and biscuits—and stirs real food into the study of America's metaphorical melting pot.

CHAPTER 1

The Burden of Bread

Bread Before Baking Powder

Who then shall make our bread? . . . It is the wife, the mother only—she who loves her husband and her children as woman ought to love.

—Rev. Sylvester Graham, *A Treatise on Bread and Bread-Making*, 1837

A woman should be ashamed to have poor bread, far more so, than to speak bad grammar, or to have a dress out of fashion.

—Catharine Beecher, *Miss Beecher's Domestic Receipt-Book*, 1858

In the first half of the nineteenth century, the future of the new nation, it seemed, was in the hands of the women who kneaded its daily bread. During the religious revival of the Second Great Awakening, bread became the focus of morality in the American home. Good bread was the measure of a good woman, a good wife, and a good mother. However, new technologies and new foods conflicted with traditional belief systems and created a moral crisis with women and bread at the center. Just as flour and warm water absorb organisms from the air to create bread, the discourse swirling around bread reflected the tensions in American life at the beginning of the nineteenth century. The concerns were many. During the first American Industrial Revolution, the daughters of Yankee farmers who lived away from home for the first time and worked in New England textile mills instead of in their family kitchens disrupted domestic functions and the mother-daughter transmission of culinary knowledge. The transportation and communications revolutions of Jacksonian America brought wheat from the heartland, and the new market economy made it possible to purchase bread more cheaply. With no history of male-controlled professional bakers' guilds or male court chefs in the United States, cooking developed in the home. American women adapted the ingredients available to them in the new country and created new foods that contributed to a uniquely American cuisine and a national identity. American women created American cuisine.

Bread was crucial to the British and the American diets and identity. A family of four or five would consume approximately twenty-eight pounds of bread each week, about one pound per person per day.[1] Bread was not just newly baked loaves eaten sliced. Fresh or stale, some form of bread was consumed at every meal of the day. Sometimes it *was* the meal. It was the mainstay of the diet of children. Bread was never thrown away. It was also used to thicken sauces, to make stuffings and puddings, and to add substance to soups and stews. Stale bread was grated into crumbs for coating oysters, chicken, and other foods before they were fried. Toast was soaked in milk or water to feed invalids. "Brewis" was the thick liquid made from crusts of soaked brown bread. "Charlotte" was fruit baked with slices of bread and butter.[2] "Brown Betty" or "pandowdy" was fruit layered with bread crumbs and butter and then baked. Bread was also a purifier, added to foods like butter to absorb objectionable odors and flavors. Bread crusts were even used to scrub walls.[3]

Women have been connected to bread making since antiquity. Words for bread, like the English language itself, come from two sources. The word "bread" is from German *brot*; the word "pantry," the place where bread was stored, is from Latin *panis*. The Old English word for "loaf," the staple of life, was *hlaf*. "Loaf-keeper," *hlaford*, became "lord"; loaf kneader, *hlaefdige*, became "lady."[4] The first law to regulate food in England was the Assize of Bread, in 1266–1267.[5]

Bread beliefs were based on ancient humoral theory, which divided foods into four humors: hot or cold, wet or dry. Anything cold and dry could lead to physical illness, a melancholy temperament, and death. Humoral theory was class-based. Suitable for laborers but unsuitable for the delicate constitutions of the upper classes were "Browne bread made of the coarsest of Wheat flower [*sic*] having in it much branne"; rye bread, which was "heavy and hard to digest"; and hard crusts, which "doe engender . . . choller, and melancholy humours."[6] Therefore, the ideal loaf of upper-class British bread was white, light, and soft.[7] Country folk had to make do with a half-rye, half-wheat loaf called "maslin."

British bread developed differently from French bread. Prized for their hard crusts, French breads were leavened with a starter that was obtained by combining flour and warm water and capturing wild yeast and bacteria from the air. Dough was saved from the previous batch of bread and continuously fed with more flour. As time progressed, this dough became increasingly sour. British bread, however, was not sour, because it was leavened with yeast that was the by-product of brewing a British beverage, ale.[8] The yeast came in two forms: barm, the foam from the top of the ale, and emptyings, or "emptins," the dregs after brewing. Beginning in the Middle Ages, British ale contained hops, so both the ale and the bread were bitter.

British colonists brought these bread concepts and technology with them; however, the exigencies of survival in a new climate and geography soon forced them to diverge from tradition. Without commercial breweries or bakeries, women had to make their own ale, then use the emptins to leaven bread. Or they had to make their own yeast.

No one knew what yeast was or how it functioned until Louis Pasteur observed it under a microscope. He published his preliminary results in 1857, followed in 1860 by his book *Mémoire sur la fermentation alcoolique*.[9] Yeast is a living organism, a single-celled fungus. When it reacts with the gluten in wheat flour, its metabolism creates carbon dioxide, a gas that leavens. Like the human body, yeast has a narrow range of temperatures in which it can thrive.

Reliable commercial yeast was not available in the United States until the Austrian Fleischmann brothers introduced it at the Philadelphia Exposition in 1876. Until then, brewer's or distiller's yeast and homemade yeast varied widely. One could not be substituted for the other. Recipes repeatedly gave alternate amounts for different yeasts or came with caveats: "Two wine-glasses of the best brewer's yeast, or three of good home-made yeast,"[10] and "Half a pint of best brewer's yeast; or more, if the yeast is not very strong."[11] An increase in yeast meant an increase in bitterness. One cookbook writer informed readers: "Many people are not aware of the difference between brewer's and other yeasts, such as distiller's. . . . Half a teacupful of brewer's yeast is as much in effect as a pint or even a quart of distiller's. . . . Distiller's yeast is always meant if not contradicted."[12] "Distiller's yeast is always meant" in *that particular* cookbook. Other cookbooks used other types of yeast. Consistency was lacking.

Catharine Beecher's *Domestic Receipt-Book* (1858) contains two pages of recipes for making yeast for bread. There is plain "Yeast," "Potato Yeast," "Home-made Yeast, which will keep Good a Month," "Home-brewed Yeast more easily made," "Hard Yeast," "Rubs, or Flour Hard Yeast," and "Milk Yeast" ("Bread soon spoils made of this").[13] She has an additional recipe in the cake section for a different kind of potato yeast, to use when making wedding cake.[14] Beecher preferred potato yeast for two important reasons: "it raises bread quicker than common home-brewed yeast, and, best of all, never imparts the sharp, disagreeable yeast taste to bread or cake, often given by hop yeast."

Yeast was time-consuming to make, and any small slip in cleanliness in the equipment used to make or store it could cause bacterial growth that rendered it useless. Beecher's instructions for keeping yeast are an additional half a page, including instructions for cleaning the container in which the yeast is stored: "Yeast must be kept in stone, or glass, with a tight cork, and the thing in which it is kept should often be scalded, and then warm water with a half teaspoonful of saleratus [baking soda] be put in it, to stand a while. Then rinse it with cold water."[15] It was also difficult to keep yeast in hot weather. At the other extreme, homemade yeast cakes that were hard enough to store had to be soaked for hours to soften and restore.

With yeast so difficult to make, and with so many variables, it is easy to understand why many American women preferred the alternative, emptins. Amelia Simmons provides a recipe for emptins in *American Cookery*, in 1796:

EMPTINS

Take a handful of hops and about three quarts of water, let it boil about 15 minutes, then make a thickening as you do for starch, which add when hot; strain the liquor, when cold put a little emptins to work it; it will keep in bottles well corked five or six weeks.[16]

Three quarts of water might seem like it would make a great deal of emptins, but emptins was liquid, not concentrated, so it took a great deal of it to leaven breadstuffs. Recipes call for it by the pint. The hops made emptins bitter, too.

It was also difficult to leaven bread because wheat was in short supply. New England was at the northern geographic and climate limit for growing wheat, the staple of British bread and not native to the Americas. This climate made wheat susceptible to diseases like Hessian fly and black stem rust.[17] The scarcity or complete lack of wheat forced New England housewives to become more creative in their use of grains and to expand the definition of bread. Corn, native to the Americas and easy to grow, was the solution.

The British colonists accepted corn quickly. They trusted their own experience and what they learned from Native Americans and ignored European scientists. In 1636 the British herbalist John Gerard wrote that corn was hard to digest, provided no nourishment, and was food for "the barbarous Indians" and "a convenient food for swine."[18] Yet in 1662 John Winthrop Jr., the governor of Connecticut, spoke to the Royal Society in London and made a case for corn as a wheat substitute, describing how to mill and prepare it: boiled, roasted, and baked. The society was not swayed.[19]

In *A Revolution in Eating*, James E. McWilliams presents this incident to bolster his Jeffersonian argument that "the way they [the colonists] thought about food was integral to the way they thought about politics"—in other words, land ownership led to republicanism.[20] This incident shows something else, too. Winthrop did not spend hours in the kitchen perfecting corn recipes. In addition to his own botanical observations, in all probability he got this information from his wife, Elizabeth Reade Winthrop, and other women in the Connecticut colony, who in turn learned it from Native American women. This shows the resourcefulness and ingenuity of colonial women and demonstrates that men, or at least Winthrop, valued their wives' labor. This kind of pragmatism and empiricism connected to food, and the ability to trust their own observations and experience, showed independent thinking and the willingness to take risks and contravene dogma from the beginning in America.

It also shows American exceptionalism through food. The colonists substituted corn for wheat in bread recipes. The British country loaf, half rye and half wheat, became half rye and half corn meal. The colonists called it "rye 'n' injun." According to humoral theory, this was not even human food. It was a heavy, dense bread, because corn does not contain gluten. Rye does, but it is of poorer quality and lesser quantity than the gluten found in wheat. The proportions of rye 'n' injun could be half and half, or two-thirds to one-third. If wheat flour was available, it was added sparingly, usually one-third of the total grains. This made the classic New England "thirded bread." Even with the addition of yeast, this dough is difficult to work. It never attains the shiny elasticity of bread made with wheat only. Even modern commercial yeast struggles mightily to lighten this bread.

One tradition the colonists kept was the connection of bread to religion. In New England, scarce wheat was prized, its use limited to "the sacrament and company."[21] This deepened the pressures on women. The ancient religious association to bread was critical, but also failing to create good bread with the little wheat that was available would have been wasteful and shameful.

American wheat presented a further obstacle for housewives because American flour was different from British flour. British writer Maria Eliza Rundell, writing in 1807, said that American flour "required almost twice as much water" as English flour to make bread. Fourteen pounds of American flour would make 21½ pounds of bread, while the same amount of English flour made only 18½ pounds.[22] This meant that in America, women had to reinvent the bread recipes they brought with them from England or the recipes available to them in British cookbooks until American women started to write their own cookbooks at the end of the eighteenth century.

Although women had always been connected to bread making, tensions about women and breadstuffs had not always existed. Cookbooks published in England and in America before the beginning of the nineteenth century contained few or no bread recipes. Recipes were unnecessary because in England the government controlled commercial bread price and quality; in America bread was baked in the home. Mothers taught daughters; each family had its own recipes.

However, two domestic sources, manuscript cookbooks and diaries, provide information about what kind of bread was baked and with what frequency. *A Booke of Cookery*, started in the seventeenth century, came to Martha Washington in the eighteenth and went on to her granddaughter to use in the nineteenth.[23] These handwritten cookbooks were treasured heirlooms, handed down from generation to generation along with the family Bible.

Another Martha, Maine midwife Martha Ballard, kept a diary for twenty-seven years, from January 25, 1785, until the last entry on May 7, 1812. She recorded how often she baked and what she baked. She wrote "we Baked" or similar entries in her diary 362 times. Sometimes she baked alone, sometimes with her daughters, sometimes with hired neighborhood girls. Sometimes the girls did the baking by themselves. Occasionally neighbors came to bake in Martha's oven.[24] Martha's diary contains 109 entries for "bread." She differentiates among "wheat," "flower" (or "flour") bread, and brown ("broun") bread (51 entries). On February 24, 1785, Martha Ballard wrote that she "Bakt & Brewd." Of the 93 entries in the diary about brewing, 24 are connected to baking. Martha simply records what she made; there is no self-consciousness about whether she is doing it right, or often enough, or if the results are good enough.

However, by the beginning of the nineteenth century the moral pressures on women to produce good bread were intense. They came from multiple sources of male authority: the pulpit, the medical profession, and academia. Three of the leading proselytizers were from the Connecticut River Valley, wellspring of the First Great Awakening and stronghold of the Second: Rev. Sylvester Graham, Dr. William Alcott, and Professor Edward Hitchcock.

In his *A Treatise on Bread, and Bread-making* in 1837, Reverend Graham preached that refined white flour was a sign of man's fall from his wholesome natural state to an artificial one, and advocated flour made from coarse ground whole wheat, heavy in bran. This flour still bears his name. At the time, it led Ralph Waldo Emerson to refer to Graham as "the prophet of bran bread and pumpkins."[25]

Graham also firmly believed that no commercial baker could make real bread; it was a quasi-religious calling reserved for women. When he wrote about food, his language was religious and sexual: "pure virgin soil" versus "depraved appetites." He also campaigned against the evils of masturbation and excessive sex in marriage. Followers of his philosophy and his food lived in Graham boardinghouses in New York and Boston.[26] They also spread Graham's teachings to the Midwest, at Oberlin College, to health spas in upstate New York, and eventually to Battle Creek, Michigan, where they were instrumental in creating breakfast cereals.

Dr. William Alcott went further than Graham. Alcott believed that no bread should be leavened with yeast, because yeast was fermented, and fermentation equaled putrefaction. Yeast was already decaying and would damage the body. Alcott had to overcome his own aversion to unleavened, unbolted (unsifted), unsalted bread: "It appeared to me not merely tasteless and insipid, like bran and sawdust, but positively disgusting." It was six months before he could force himself to become accustomed to this bread.[27] Alcott echoed Graham as to who should bake the bread: the woman who has "a love for her husband and family," because "no true mother, daughter or sister . . . can long remain ignorant of bread-making."[28]

Edward Hitchcock, professor of chemistry at Amherst College, believed that eating sparingly was the key to longevity, a philosophy also espoused by Nathan Pritikin in the late twentieth century. Hitchcock cites examples of men who thrived on a Spartan diet and lived well past one hundred years of age.[29] Hitchcock's diet regimen consisted of twelve ounces of solid food and twenty of liquid per day. The solid food was heavy on bread.

Breakfast at 7:00 a.m.

Stale bread, dry toast, or plain biscuit, no butter	3 ounces
Black tea with milk and a little sugar	6 ounces

Luncheon at 11:00 a.m.

An egg slightly boiled with a thin slice of bread and butter	3 ounces
Toast and water	3 ounces

Dinner at 2:30 p.m.

Venison, mutton, lamb, chicken, or game, roasted or boiled	3 ounces
Toast and water, or soda	1 ounce
White wine, or genuine Claret, one small glass full	1 ounce

Tea at 7:00 p.m. or 8:00 p.m.

Stale bread, biscuit, or dry toast, with very little butter	2 ounces
Tea (black) with milk and a little sugar	6 ounces[30]

Hitchcock went further than Graham and Alcott in his attitude toward women and bread. Under the heading "How far the blame is to be imputed to females," Hitchcock condemns women for making food that is expensive and will kill their loved ones instead of making good, wholesome bread. He blames men, too, but he blames women first, for two and a half pages. His case against men is only one page.[31]

All three of these men wrote and lectured extensively. Their writings were in the tradition of upper-class male-to-male lifestyle manuals that began in antiquity. Men have always looked to food for immortality, and although those early books contained dietary advice, they did not contain recipes until the Renaissance. In the Catholic countries, cookbooks were written by professional male chefs for the upper classes and contained recipes and advice about the court and its cuisine. The first cookbook written by a Catholic woman did not appear until after the French Revolution. Mme. Mérigot's *La Cuisiniere Républicaine*, from 1794 or 1795, was a small pamphlet that consisted of thirty-one potato recipes, most of them one or two sentences long.[32]

On the other hand, American cookbooks written by women come from a more recent tradition of female literacy that began with the Protestant Reformation in the sixteenth century. Martin Luther said, "God doesn't care what you eat," and broke with the feast and fast days of the Catholic Church. With its emphasis on direct communication with the creator and personal knowledge of the Bible, Protestantism emphasized literacy, including female literacy. In the early modern period, upper-class men began writing manuals to instruct women and included recipes.[33] By the second half of the seventeenth century, middle-class women in northern European Protestant countries were writing cookbooks. In 1667 *The Sensible Cook* was published in the Netherlands. Food historians believe that the anonymous author was a woman. In Britain in 1673, Hannah Woolley wrote *The Gentlewoman's Companion*. In eighteenth-century England, Hannah Glasse, Susannah Carter, and E. Smith wrote comprehensive cookbooks that explained how to cook every type of food using sophisticated culinary techniques. The books also became popular in colonial America.

In the early republic, American women began to write cookbooks and books of household management themselves. These were also statements of philosophy, female-to-female communication. They spread rapidly because there was 100 percent literacy among women in New England by 1840.[34]

Historians have not analyzed early American cookbooks in depth, and not with an eye to breadstuffs and chemical leavenings. Three books written in the 1980s by women historians began to examine women's work in the home. In *Never Done*, her groundbreaking 1982 history of housework, Susan Strasser examined cooking in American households. However, she used only three cookbooks, by Sara Josepha Hale, Mary Randolph, and Catharine Beecher. The majority of the other information came from secondary sources written by men.[35] In *More Work for Mother: The Ironies of Household Technology from the Open Hearth to the Microwave* (1983), Ruth Schwartz Cowan analyzed diaries, letters, and probate records to find out what housework was like for women in America. However, she did not use cookbooks.[36] Glenna Matthews's 1987 *"Just a Housewife"* used cookbooks by Amelia Simmons, Randolph, Beecher, and unnamed others from the Schlesinger Library at Radcliffe and a private collection.[37] In *Eat My Words* in 2003, Janet Theophano examined cookbooks more closely, but from an anthropological perspective.[38]

Many of the women who wrote early cookbooks were what historian Natalie Zemon Davis calls "women on the margins": orphans, widows, single women, or married women whose husbands for whatever reasons could not support the family.[39] Amelia Simmons, who wrote the first American cookbook, *American Cookery*, in 1796, identifies herself as "an American orphan."[40] Wealthy Mary Randolph, from one of the First Families of Virginia, was the mistress of a tobacco plantation with forty servants. In 1808, when political factors caused her family to lose their money, she opened a boardinghouse. In 1824 she wrote her cookbook, *The Virginia Housewife*.[41] Eliza Leslie and her mother opened a boardinghouse when Eliza's father died and they were desperate. In 1827 Eliza wrote *Seventy-five Receipts for*

Pastry, Cakes, and Sweetmeats.[42] Lydia Maria Child wrote *The American Frugal Housewife* in 1829 to support her family, because her husband was in prison.[43] Catharine Beecher, one of the most influential domestic educators of the mid-nineteenth century, author of *A Treatise on Domestic Economy* (1841) and *Miss Beecher's Domestic Receipt-Book* (1846), never married after her fiancé died at sea.[44] Like a nineteenth-century Phyllis Schlafly, Beecher spent her life traveling, speaking, writing, and telling women why they should stay home and bake bread.

American women cookbook writers were conscious of their identity as free people in a new country. They repeatedly referred to themselves and their cooking as American. They were outspoken about their vision for the new country and the place of food and female cooks in it. *The Cook Not Mad* contained recipes for "good *republican dishes*," not recipes of "English, French and Italian methods of rendering things indigestible."[45] The book was published "in the fifty-fifth year of the Independence of the United States of America, A.D. 1830."[46] In the new republic, Election Day was a holiday and cause for celebrating. Simmons has a recipe for "Election Cake"; Beecher has "Hartford Election Cake." On May 27, 1807, Martha Ballard wrote in her diary, "I made a Cake for Supper. it is Election day."[47] Simmons also has "Independence Cake" and a "Federal Pan Cake" of rye, cornmeal, salt, and milk fried in lard.[48] Sarah Josepha Hale, founder of the Boston-based *Ladies' Magazine* and daughter of a Revolutionary War officer, echoed Sam Adams when she organized a Committee of Correspondence to raise money for the Bunker Hill Monument.[49]

Women who wrote cookbooks had to address the teachings of the popular Graham, Alcott, and Hitchcock. Hale would have none of them. Neither would Child. Hale considered bread making so important that she began *The Good Housekeeper* with it. She did include a recipe for "Brown or Dyspepsia Bread," but grudgingly.

"Dyspepsia" was a catchall term for any kind of indigestion in the nineteenth century. However, Hale refused to call it graham bread: "This bread is now best known as Graham bread—not that Doctor Graham invented or discovered the manner of its preparation, but that he has been unwearied and successful in recommending it to the public." She says there is nothing wrong with the bread, "though not to the exclusion of fine bread." Her recipe undercut Graham by telling readers to sift the flour first. This, of course, removed the fibrous particles that Graham believed provided the benefit.

BROWN OR DYSPEPSIA BREAD

Take six quarts of this wheat meal, one tea-cup of good yeast, and half a tea-cup of molasses, mix these with a pint of milk-warm water and a tea-spoonful of pearlash or sal aeratus [*sic*].[50]

Child, too, writing in 1832, had a recipe for dyspepsia bread, which she attributed to the *American Farmer* periodical. The ingredients were the same as in Hale's recipe but in different proportions. The unintended effect of dyspepsia bread was that Americans became habituated to sweeteners in their daily bread, which had previously been unsweetened, and to thinking of sugar and molasses as healthy. Child also mentions that "Dyspepsia crackers can be made with unbolted [unsifted] flour, water and saleratus"—the beginning of graham crackers.[51] In the twenty-first century, the first ingredient in commercial graham crackers is sugar.

Even if they disagreed with Graham and his followers, women had other moral issues connected to bread. In *The American Frugal Housewife* in 1844, Lydia Maria Child exhorted women to "make your own bread and cake" and stopped just short of accusing women who did not bake their own bread of being guilty of one of the seven deadly sins, sloth: "your domestic or yourself, may just as well employ your own time, as to pay [commercial bakers] for theirs." Sarah Josepha Hale (1841) did speak of food consumption in terms of "sin." In this category she included eating anything immoderately.[52] The underlying issue: "there is great danger of excess in all indulgences of the appetites," making the unspoken connection between food and sex.[53]

In two books, Catharine Beecher also made bread baking the moral measure of a woman. *Miss Beecher's Receipt-Book* was a supplement to her *Treatise on Domestic Economy*. This manual on "every aspect of domestic life from the building of a house to the setting of a table" was extremely influential and was even on the curriculum at Ralph Waldo Emerson's school.[54] Both of Beecher's books provided the platform by which she promulgated what her biographer, Kathryn Kish Sklar, calls Beecher's "evangelical leadership to meet the threat of national wickedness and corruption."[55] Bread baking was at the forefront of the crusade.

The connections between cooking and morality were not only in cookbooks. They also appeared in nineteenth-century American novels. Mark McWilliams found that "for novelists as varied as Fanny Fern, Caroline Howard Gilman, Nathaniel Hawthorne, Augustus Baldwin Longstreet, Harriet Beecher Stowe, and Susan Warner, women who cook well serve as moral exemplars while women who cannot face social stigma."[56] However, cooking well was not enough. An additional burden was that women had to bake bread, which was strenuous manual labor, time-consuming, and messy, and they had to make it look effortless. Feminist scholar Margaret Beetham points out that in the nineteenth century, "it was an important part of the masculine fantasy of the domestic [middle-class femininity] that the work of maintenance was invisible."[57]

This was not solely a male fantasy. Women internalized it. In *The Minister's Wooing*, her 1859 novel, Harriet Beecher Stowe wrote: "The kitchen of a New England matron was her throne room, her pride; it was the habit of her life to produce the greatest possible results there with the slightest possible discomposure."[58] Her main character personified the "angel of the hearth," whose four virtues were piety, purity, domesticity, and obedience. This Victorian idealized image of the housewife as unflappable and perfectly groomed endures. It was popularized in the 1950s and 1960s on television shows such as *The Adventures of Ozzie and Harriet* (1952–1966), *Father Knows Best* (1954–1960), and *Leave It to Beaver* (1957–1963). The epitome of this ideal, which persists into the twenty-first century, is Martha Stewart.

In the growing market economy, there were also serious financial pressures on women to produce good bread, especially women who ran boardinghouses. All of the food, including the bread, was notoriously poor in these establishments. Mary Randolph's was an exception. As Wendy Gamber points out in her study of boardinghouses in the nineteenth century, new technology created new expectations. After cooking shifted from open hearth to stoves, which supposedly made baking easier, women were expected to bake yeast bread more often. As white yeast bread came to signify middle-class status, boardinghouse residents complained when they got breadstuffs leavened instead with "soda, saleratus, and cream of tartar."[59] In the novel *Ten Dollars Enough*, Mrs. Bishop, a boardinghouse resident and the main character, yearns for freshly baked bread instead of chemically leavened biscuits.[60]

In urban areas, bread began to be produced outside the home in small, one-man, one-oven bakeries.[61] With commercial bread available in cities and towns, daughters had no incentive to learn how to make it. By 1857, Philadelphia had ninety-nine bakeries. Even in the more recently settled West, cities like Cincinnati, Cleveland, Detroit, and Milwaukee all had multiple commercial bakeries.[62] Still, by the 1850s more than 90 percent of bread was made in the home, so most women bore the burden of bread baking.[63]

Commercial bakeries contributed to the disruption of the transmission of mother-daughter knowledge. Another sign of this disruption is that lengthy recipes explaining how to make bread began to appear in cookbooks, such as Catharine Beecher's. Even basic bread, always made in bulk, was extremely time-consuming because of the volume.

WHEAT BREAD OF HOME-BREWED YEAST

Sift eight quarts of flour into the kneading tray, make a deep hole in the middle, pour into it a pint of yeast, mixed with a pint of lukewarm water, and then work up this with the surrounding flour, till it makes a thick batter. Then scatter a handful of flour over this batter, lay a warm cloth over the whole, and set it in a warm place. This is called sponge.

When the sponge is risen so as to make cracks in the flour over it (which will be in from three to five hours), then scatter over it two tablespoonfuls of salt, and put in about two quarts of wetting, warm, but not hot enough to scald the yeast, and sufficient to wet it. Be careful not to put in too much of the wetting at once.[64]

> Knead the whole thoroughly for as much as half an hour, then form it into a round mass, scatter a little flour over it, cover it, and set it to rise in a warm place. It usually will take about one quart of wetting to four quarts of flour. [This second rising also took hours.]
>
> In winter, it is best to put the bread in sponge over night, when it must be kept warm all night. In summer it can be put in sponge early in the morning, for if made over night, it would become sour.[65]

Vagaries of yeast potency and weather were compounded by wide variations in ovens. Before the Civil War, ovens were brick, built into the wall next to the fireplace. Preparing the oven for baking was a lengthy process. The baker had to build a fire and let it cool down. Then she had to sweep the floor of the oven clean of ashes and check to make sure the temperature was right. Thermometers had been invented by Fahrenheit and Celsius in 1714 and 1742, respectively, but did not come into widespread use in home ovens until the twentieth century.[66] Until then, the only way to determine the temperature of the oven was to put something in it and observe what happened. Most cookbooks came with instructions on how to do this. In 1857 Eliza Leslie instructed her readers to use the twenty-second rule: "Try the heat of the oven by previously throwing in a little flour. If it browns well, and you can hold your hand in the heat while you count twenty, it is a good temperature for bread. If the flour scorches black the oven is too hot, so leave the oven open a little while [until] it becomes cooler."[67] The following year, Catharine Beecher, under the heading "How to know when an Oven is at the right Heat" advised her readers, too, to use the twenty-second rule but also pointed out its shortcomings: "These . . . are not very accurate tests, as the power to bear heat is so diverse in different persons; but they are as good rules as can be given, where there has been no experience."[68]

If baking bread was a burden, housewives could still find a way to pursue happiness in the women's sphere of the home: they could bake other things. Even though eighteenth-century cookbooks did not contain many bread recipes, recipes for cakes abound. Pre–baking powder cakes were of three basic types: yeast, sponge, and pound. Yeast-risen cakes were the most common. They were like bread but more luxurious. Daily bread was made of three or four ingredients: grain, water, yeast, and sometimes salt. Yeast-risen cakes, however, included expensive ingredients that daily bread did not: eggs, sugar, milk, butter, spices, fruits, nuts, and alcohol.

These American cakes were in the tradition of the British Great Cake, which in turn was like Italian *panettone,* which means "big bread." As Karen Hess has pointed out, these cakes were probably also baked like bread: "cast in large round loaves directly on the floor of the brick oven," not in pans, molds, or baking hoops.[69] "Cast" means shaped into a round and placed on the bottom of the oven. For sponge cakes, the egg yolks and whites were separated, and the whites were beaten to leaven the cake. This was not easy to do: "The want of butter renders it difficult to get light."[70] Pound cake was so named because it contained a pound each of butter, sugar, and flour, plus a dozen eggs.

Making these cakes was even more laborious than making bread. In addition to yeast, the ingredients required lengthy preparation: "In preparing cake, the flour should be dried before the fire, sifted and weighed; currants washed and dried; raisins stoned; sugar pounded, and rolled fine and sifted; and all spices, after being well dried at the fire, pounded and sifted."[71] Then the batter either had to be kneaded, for yeast-risen cake, or the egg whites beaten for sponge cakes. Hickory rods were preferred to wire egg beaters.[72] Pound cake was, and still is, dense, because it contains no leavener.

Butter and sugar had to be "creamed"—worked together with the hands until they were the consistency of cream. Then after the other ingredients were added, instructions were "Beat it all well together for an hour."[73] This was not unusual. Eliza Leslie advised, "Have this done by a man servant."[74] As a result, cakes were reserved for important occasions.

However, in the first half of the nineteenth century, a revolution occurred in the preparation of breadstuffs. Once they escaped the limitations of yeast, American women were free to experiment, innovate, and create new kinds of food, removed from religion. This culinary revolution started with cake.

CHAPTER 2

The Liberation of Cake

Chemical Independence, 1796

Six pound flour, 2 pound honey, 1 pound sugar, 2 ounces cinnamon, 1 ounce ginger, a little orange peel, 2 teaspoons pearl-ash, 6 eggs; dissolve the pearl-ash in milk, put the whole together, moisten with milk if necessary, bake 20 minutes.

—Recipe for Honey Cake, Amelia Simmons, *American Cookery*, 1796

Cake of every sort is to be partaken of as a luxury, not eaten for a full meal. Those who attend evening parties several times in a week, can hardly take too small a quantity of the sweet and rich preparations.

—Sarah Josepha Hale, *Early American Cookery: The "Good Housekeeper,"* 1841

The most prominent word on the cover of *American Cookery*, the first cookbook written in the United States, in 1796, is "CAKES." It is centered on the cover, top to bottom and left to right. The letters are all uppercase; spaces between them make the word seem even larger. The cover makes no mention of bread; American cookery, evidently, was all about cake. Revolutionary new chemical leaveners made cake convenient, cheap, quick, and easy to bake. American women experimented with these leaveners, which shifted cakes, muffins, biscuits, cookies, and other sweet breadstuffs from luxuries to everyday foods. Chemical leavening created an entire new category of food, with new textures and flavors. A concomitant increase in infrastructure and in flour milling made finely ground wheat available. These new foods, in turn, became the occasion for new types of social events. By the second quarter of the nineteenth century, cake had become so common that Sarah Josepha Hale had to warn against overindulging in it.

In 1796 a new leavener appeared in print for the first time. It created a revolution in baking as profound as the discovery of yeast thousands of years earlier.

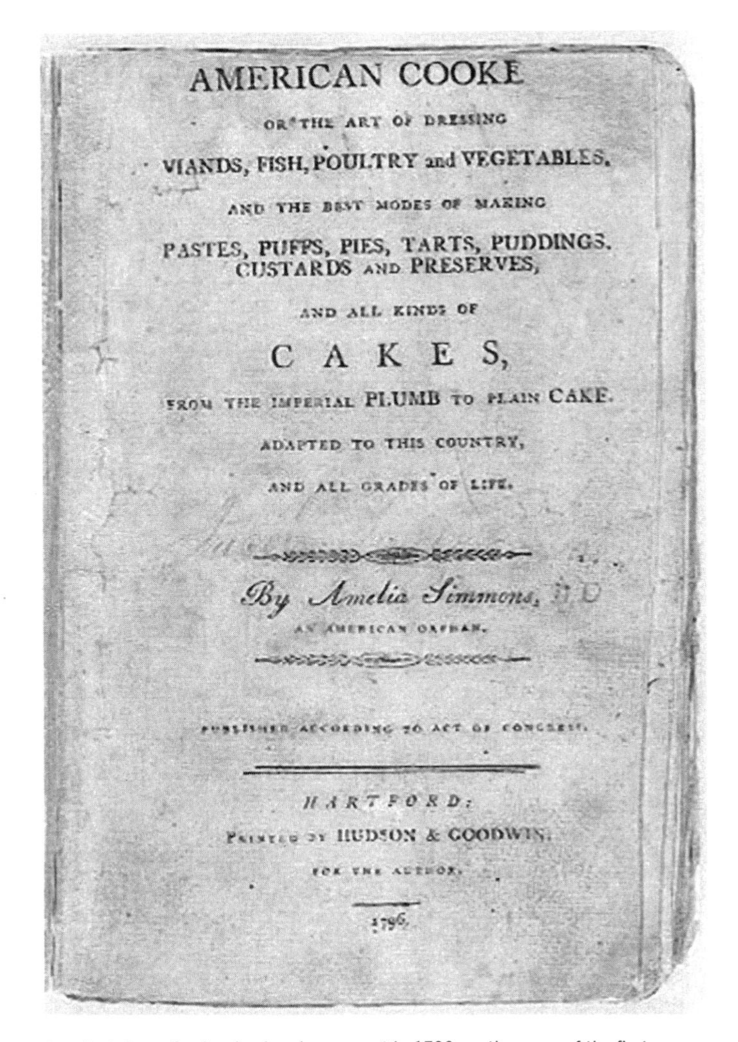

AMERICAN COOKE

OR THE ART OF DRESSING

VIANDS, FISH, POULTRY and VEGETABLES,

AND THE BEST MODES OF MAKING

PASTES, PUFFS, PIES, TARTS, PUDDINGS,
CUSTARDS AND PRESERVES,

AND ALL KINDS OF

CAKES,

FROM THE IMPERIAL PLUMB TO PLAIN CAKE.

ADAPTED TO THIS COUNTRY,

AND ALL GRADES OF LIFE.

By Amelia Simmons,

AN AMERICAN ORPHAN.

PUBLISHED ACCORDING TO ACT OF CONGRESS.

HARTFORD:

PRINTED BY HUDSON & GOODWIN,

FOR THE AUTHOR.

1796.

America's love of cakes is already apparent in 1796, on the cover of the first cookbook written by an American. American Treasures of the Library of Congress.

This leavener was first written about by Amelia Simmons in *American Cookery*, published in Hartford, Connecticut. Seven of the recipes in *American Cookery* contain a new ingredient, a baking shortcut called pearlash, sometimes spelled "pearl ash" or "pearl-ash." This was a mild alkali, potassium carbonate.[1] Simmons's use of pearlash is casual; she makes no mention of it being new, out of the ordinary, or difficult to acquire. Food historian Karen Hess has pointed out that "print lags behind practice, sometimes by centuries."[2] This means that pearlash was probably in widespread use in the United States by the time Simmons wrote her book.

Pearlash is a refined form of potash, a by-product of burned plant material, often wood. The earliest mention of potash as a food ingredient appeared in a Dutch professional bakers' cookbook published in the Netherlands in 1753, in a recipe for pretzels.[3] Perhaps Dutch bakers in New York's Hudson River Valley and in northwestern Connecticut spread this knowledge to American women bakers, who further refined the potash to pearlash. In any case, it was pearlash, and the new types of baked goods made with it, that became popular in American cookbooks.

The first patent issued in the United States was for pearlash. On April 10, 1790, Congress passed the Patent Act in order "to promote the Progress of useful Arts."[4] On July 31, 1790, the first patent was issued to Samuel Hopkins of Vermont. It was for "an improved method of making pure pearl ash from wood ashes."[5] The U.S. Patent Office website says that this was a patent for potash, which was used in fertilizer and making soap. The website completely ignores pearlash, which was listed first and repeatedly in the patent and was the predominant part of it. The use of pearlash as a leavener in baked goods was new. With so many forests available to burn, in 1792 the United States exported eight thousand tons of pearlash to Europe.[6]

Preparing pearlash was time consuming, but housewives could prepare it in a cast-iron kettle. "Lye, crudely leached from their fire-ashes, was slowly boiled down to 'salts.' Then, with much stirring, this was 'scorched' to burn out bits of vegetable matter. This was accomplished . . . at a temperature well below red heat. . . . This scorching produced a poor grade of pearl ash, granular in texture and mottled yellow to gray in color. It was often the very first product of the pioneer housewife on the backwoods frontier."[7]

Ashes have been used beneficially in food for millennia. In ancient Rome, Cato wrote recipes for curing olives with lye, which is made of ashes. Pre-Columbian Mesoamericans nixtamalized corn with ashes to render it more pliable and imbue it with B vitamins, which corn lacks. Lye—*lut*—gives *lutefisk*, the Scandinavian fish dish, its name. In cheese making, ashes preserve and demarcate. At the end of the first day's processing, a layer of vegetable ash was sometimes spread over the cheese as a barrier to the air. This dark line running horizontally through cheese has been visible in French Morbier cheese since the nineteenth century and is in Cypress Grove, California's, prizewinning Humboldt Fog in the twentieth. Martha Ballard mentions potash and "Pott ash" a total of three times in her diary, once to say that she bought some. What she did with it is unrecorded. Ashes were also considered healthy. In the "Simple Remedies" section at the beginning of *The American Frugal Housewife*, Lydia Maria Child advises her readers, "A spoonful of ashes stirred in cider is good to prevent sickness at the stomach. Physicians frequently order it in cases of cholera-morbus."[8]

Pearlash created new types of food and food with new textures. Simmons has thirty recipes for leavened cake and bread: fifteen of them call for emptins, seven for yeast, and six for pearlash. One bread recipe uses both pearlash and yeast. Simmons usually says to dissolve the pearlash in milk or water before adding it to the other ingredients.

Pearlash is the leavening in the first written recipe using the word "cookies," in *American Cookery*. Simmons also has three recipes for a new style of American gingerbread cookie, all leavened with pearlash. What Europeans call gingerbread, Americans call "gingerbread people," usually made at Christmas with dough rolled out and cut into the shapes of gingerbread boys and girls. The European version is extremely crisp, like a cracker, but Americans prefer their gingerbread people softer. Simmons also has the first recipe for what Americans know as gingerbread, "Soft Gingerbread to be baked in pans."[9]

Simmons was unsatisfied with the first edition of her cookbook and came out with a second edition later in 1796, published in Albany, New York. She deleted one cake recipe and added eleven new ones. The total number of cake recipes increased from thirty-six to forty-six. The cake section, as the cover promised, contained more recipes than any other section of the book by far. *American Cookery* was popular: "printed, reprinted and pirated for 30 years after its first appearance"[10] (see appendix table A-1).

Amelia Simmons's use of pearlash in *American Cookery* caused a sensation when it reached England.[11] Pearlash soon appeared in other American cookbooks. In *The Cook Not Mad*, published in 1831, chemical leavening is in the titles of two cakes: "Pearlash Cake" and "Mineral Cake." More importantly, this cookbook does not limit the use of pearlash to cakes or cookies. It expands to include other types of foods, such as "Common Pancakes" (Recipe 150), "Shortcakes" (Recipe 165), and "Buckwheat Cakes" (Recipe 172). Like *American Cookery*, this book also has recipes for cookies, such as "Jackson Jumbles," in honor of President Andrew Jackson (Recipe 166). However, "Biscuits," "Tea Biscuits," and "Butter Biscuits" were still made with yeast or emptins: "A quart of milk, one pound of butter, gill of yeast, made sufficiently hard to rise with flour" (Recipe 123). "Muffins" then were what Americans now call English muffins, rounds of yeast-risen dough that could be cooked on a griddle top, in the interest of speed: "One quart of milk, four eggs, small cup of butter, some yeast, to be made stiffer than pound cake, bake it on a griddle in drops."[12]

Unlike yeast-risen cakes, cakes leavened with pearlash were not laborious, time-consuming, or expensive. According to Sarah Josepha Hale, a family of four or five people would need bread made of twenty-one quarts of flour each week, which would make seven loaves that weighed four pounds each and took one and a half hours to bake. She recommended baking once a week, even though in winter, bread might keep up to two weeks.[13] On the other hand, cakes and baked goods such as gingerbread that were leavened with chemicals used approximately one quart of flour, mixed easily without kneading, and baked for twenty to forty minutes. This meant that bread would be stale but cake would be fresh. In addition, women regarded this new style of American cake, not heavy with butter, as healthier. Hale sounded the alarm: "The kinds of cake most apt to prove injurious are pound cake and rich plum cake."[14] Hale advocated sponge cake or gingerbread and provided three recipes for the latter, all leavened with pearlash.[15]

As urbanization increased, family size decreased, so there was no need for the enormous great cakes that fed villages. The average family size in 1790 was 5.74 members; by 1860 it had dropped to 5.16.[16] This reflects a trend in American family demographics toward more households with fewer members, which continued into the 2010 census.

There were concomitant changes in the availability of flour. When the Erie Canal opened in 1825, ships on the canal brought wheat from upstate New York and the Midwest to the Northeast, cheaply and quickly. Shipping a ton of grain from Buffalo to New York City, which previously had taken almost three weeks, now took less than one week. The cost dropped, too, from one hundred dollars to five dollars.[17] Around Rochester, New York, wheat production doubled or tripled from 1822 to 1835.[18] So many different kinds of flour were available that Catharine Beecher instructed her readers, "When good flour is found, *notice the brand*, and seek the same next time."[19] For Eliza Leslie, that flour was from the Hiram Smith Mills in Rochester, New York.[20]

Like cake, in the early nineteenth-century biscuits could be made with yeast, chemical leaveners, or neither. Martha Ballard bought and made biscuits, which she spelled "bisquits" or "biskits." These are not what Americans today know as biscuits. The word "biscuit" is French and means "twice cooked." Its origins are the Italian "biscotti," which means the same thing. In America, biscuits were not twice cooked. The word was adopted as a name for small, shaped breadstuffs. In Martha's diary, biscuits are rare, a luxury item. Of a total of thirteen entries for "bisquit" or "biskit," ten refer to Martha buying biscuits at a store, a dozen at a time; receiving them as a gift; or administering them to the sick, along with wine. They could have been something like the soft "Naples Biscuit," what we call ladyfingers. These were made with eggs, flour, and sugar. Mary Randolph's recipe calls for a dozen eggs, a pound of flour, and a pound of sifted sugar, beaten "all together till perfectly light." They required special pans so that each cake "will be four inches long, and one and a half wide."[21] Perhaps this was equipment that Martha did not have. Or Martha might have bought crisp biscuits, like hard European gingerbread.

In addition to the store-bought biscuits, Martha also made biscuits three times. Twice she baked them in tandem with other foods: once with "cake and broun bread," once with pies. On one rare occasion, Sunday, July 16, 1786, Martha baked nothing but biscuits: "Bakt Bisquits for Jono [Jonathan, her son] to Carry to Sea."[22] These were sea biscuits, also known as hardtack. Made from only two ingredients, flour and water kneaded together, hardtack had been a staple in the British navy for centuries. It was the bane of the sailor's existence. With no fat, dairy, or eggs to make them go rancid, these biscuits had very long keeping power. Even a little dampness did not make a dent in them; they needed long soaking in the liquid of a stew or chowder to soften them. Hardtack was one of the ingredients in a sailor's meat-and-vegetable stew called lobscouse, which began with "biscuit pounded fine."[23] Often chowder was made with the biscuits layered in it so that as they softened, they also thickened the stew.

Another type of biscuit, "Beaten Biscuit," was the pride of the South. De rigueur at the dinner table, often as an accompaniment to country ham, beaten biscuits denoted hospitality and status. The biscuits had a distinctive look. They were cut into rounds and the tops were pricked with a fork. They were extremely labor-intensive, which meant slave labor. Martha Washington's *Booke of Cookery* has a recipe for "Bisket" that says, "The longer you beat it the better it will be . . . you must beat it till it will bubble."[24]

Beating incorporated air into the batter, which produced bubbles that made a popping sound when they burst. Mary Randolph's recipe for "Apoquiniminc Cakes," made with flour, butter, egg, and salt, instructs bakers to "beat it for half an hour with a pestle." Randolph's other biscuit recipes say only "when well kneaded" or "knead it well" with no time component, but these would have been time-consuming too.[25]

Northern women in antebellum America saw a moral choice between beaten biscuits and bread made with cornmeal and expressed this in their cookbooks. Four years before the Civil War began, Eliza Leslie, from Quaker abolitionist Philadelphia, included a recipe from a slave state, "Maryland Biscuit," in her cookbook. It is not just a recipe; it is a political statement that connects racism, feminism, and democracy to food. Leslie wrote at length deploring both the food itself and the labor involved in producing it.

MARYLAND BISCUIT

Take two quarts of sifted wheat flour, and add a small tea-spoonful of salt. Rub into the pan of flour a large quarter of a pound of lard, and add, gradually, warm milk enough to make a very stiff dough. Knead the lump of dough long and hard, and pound it on all sides with a rolling-pin. Divide the dough into several pieces, and knead and pound each piece separately. This must go on for two or three hours, continually kneading and pounding, otherwise it will be hard, tough, and indigestible. Then make it into small round thick biscuits, prick them with a fork, and bake them a pale brown.

This is the most laborious of cakes, and also the most unwholesome, even when made in the best manner. We do not recommend it; but there is no accounting for tastes. Children should not eat these biscuits—nor grown persons either, if they can get any other sort of bread.

> When living in a town where there are bakers, there is no excuse for making Maryland biscuit. Believe nobody that says they are not unwholesome. Yet we have heard of families, in country places, where neither the mistress nor the cook knew any other preparation of wheat bread. Better to live on indian cakes.[26]

Six pages earlier in the book, Leslie did provide a recipe for "Indian Cakes"—"the perfection of corn cakes." This, too, is a political statement: "The cook from whom this receipt was obtained, is a Southern colored woman, called Aunt Lydia."[27] Here, Leslie did what was rare in cookbooks, even into the twentieth century: she acknowledged that this recipe had been created by a black female cook and even named her. As Toni Tipton Martin points out, not only were the South's black female cooks, the generic "Aunt Dinahs," not acknowledged, but their culinary achievements were routinely claimed by the white mistress and handed down in print or in family cookbooks as her own.[28]

It is curious that this statement came from Eliza Leslie, because Lydia Maria Child was "perhaps the most important of the free white female abolitionist writers."[29] Although Child kept her abolition work and her cookbook work separate, eventually public opinion against her abolition work caused her cookbook to go out of print.

American women experimented with other chemical leavener acids, alkalis, mineral salts, and combinations. Spirit of hartshorn, so called because it was distilled from the antlers of deer, was very pungent because it was approximately 28.5 percent ammonia gas.[30] Hartshorn came in a lump and had to be ground into a powder. It was also known as baker's ammonia. This created confusion because baker's ammonia is ammonium carbonate, which has a different molecular structure.[31] To compound the confusion, some recipes called for sal volatile, more commonly known as smelling salts. Chemically, these are also ammonium carbonate.[32]

In the 1830s and 1840s, baking powders were coming into use in England too. In 1837 and 1838, respectively, two men, Dr. Whiting and Mr. Sewall, took out separate patents on a process to leaven bread using muriatic acid, which also had caustic solvent properties.[33] In London in March 1842, George Borwick & Sons, Ltd., began selling a baking powder that was "a mixture of starch and citrate powder."[34] It was sold wrapped in paper for a penny a packet. The wrapping was done by Borwick's daughter and other female children.[35] Baking powder was just one chemical product that the British pioneered during the Second Industrial Revolution but failed to capitalize on because they did not make "the essential investments in production, distribution, and management" that would have allowed their businesses to expand.[36]

In 1846 an important new leavener appeared in the United States. Church & Dwight, the company that later became Arm and Hammer, began selling baking soda, the mineral sodium bicarbonate, as a leavener for baked goods.[37] Another popular leavener was saleratus, Latin for "salt aerated." This was the popular name for potassium bicarbonate, which gained popularity in the 1850s.[38] However, it contributed to the confusion over leavenings, because sodium bicarbonate—what we now call baking soda—later was also called saleratus.[39] Saleratus, along with baking soda and pearlash, was used to bolster ineffective or sour yeast, or dough that rose too long or in too much heat in the summer.[40] Catharine Beecher also refers to a "super carbonate."[41]

Although beer and brewing by-products were used as leaveners in the colonial United States, wine was not. However, in the nineteenth century, the residue of wine making began to provide an important source of chemical leavening. Cream of tartar, also known as "cream tartar," contained the chemical argol. The United States did not have a wine industry, so the cream of tartar was imported from France and Italy. It was "best bought in lumps and then pulverized and kept corked."[42]

In 1846 the first edition of Catharine Beecher's *Domestic Receipt-Book* includes a recipe that uses both a type of baking soda and cream of tartar, an early attempt at baking powder. This recipe is the prototype of baking powder biscuits.

CREAM TARTAR BREAD

Three pints of dried flour, measured after sifting.
Two cups of milk.
Half a teaspoonful of salt.
One teaspoonful of soda (Super Carbonate).
Two teaspoonfuls of cream tartar.

Dissolve the soda in half a tea-cup of hot water, and put it with the salt into the milk. Mix the cream tartar *very* thoroughly in the flour: the whole success depends on this. Just as you are ready to bake, pour in the milk, knead it up sufficiently to mix it well, and then put it in the oven as quick as possible. Add either more flour or more wetting, if needed, to make dough to mould. Work in half a cup of butter after it is wet, and it makes good short biscuit.[43]

Another cookbook that combined cream of tartar and baking soda as a leavening agent was *Practical American Cookery and Domestic Economy*. First published in 1855 in New York, this cookbook used cream of tartar and baking soda together in several different kinds of breadstuffs, such as the crust for a veal pot pie. This is a new kind of pie crust. In the Middle Ages, rectangular inedible pie crusts called "coffins" served as container and lid for the contents. By the late eighteenth century, these old-style crusts were in cookbooks alongside flaky edible crusts.

The crust in this 1855 veal pot pie recipe was a third, new type. If it had been a chicken pie, it would be called "chicken and dumplings." This could be called "veal and dumplings."[44] The dough mixes easily, is dropped on top of the stew, and covered. The steam from the stew cooks the dough. No oven required. When biscuit dough was dropped on top of sweetened fruit and then baked, it was called a "cobbler." The name referred to the method of making the topping. Instead of rolling the dough out into one solid sheet like a pie crust, the topping was cobbled together piece by piece. This produced a more rustic look, which was also easier on the cook.

Three of the recipes in *Practical American Cookery* that combine cream of tartar and baking soda are for cake. One is for doughnuts, called "Soda Doughnuts." The ingredient was so unusual, it appears in the title. Before this time, doughnuts were yeast-risen.

SODA DOUGHNUTS

One quart of flour, one and a half cups of milk, one teaspoon full of soda, two of cream of tartar—soda dissolved in the milk, cream of tartar rubbed dry into the flour—two eggs, sugar and cinnamon to your taste. Boil in hot fat. They are nice when fresh, but will not keep long.[45]

This recipe is important because it gives instructions on how to use baking soda and cream of tartar, which indicates that these were new ingredients. Later recipes in this book, for example, "Soda Cake" and "Railroad Cake," also combine cream of tartar and baking soda but do not give instructions. *Practical American Cookery and Domestic Economy* is a compilation cookbook, so there might be several reasons for the discrepancy. Perhaps the woman who wrote the recipe for "Soda Doughnuts" is more precise than the writers of the other recipes. Or maybe she first combined these ingredients herself and is sharing knowledge she knows other women do not have.

All of these leaveners had major advantages over yeast. One is that they were not as temperamental as yeast. They did not have to be warmed up in winter; they could not over-rise and go rancid in summer. Most importantly, these leaveners reacted to moisture or heat, so they were independent of gluten; thus they could raise maize. Light, fluffy cornbread was another American culinary revolution.

Chemical leaveners also had the advantage of spontaneity. Cake making was not locked into a certain day of the week as bread baking was. Cakes were voluntary; they did not have to be baked on a certain day or at all. No advance planning was necessary. Housewives did not have to set a sponge, or make sure they had yeast, or make yeast or emptins. They could just uncork the powders in the pantry or run down to the apothecary and buy a premeasured amount of the necessary chemicals.

Chemically leavened cakes also moved housewives away from baking in large batches. Several different kinds and shapes of cake could be made in much less time than it took to make one batch of bread. Cake lent itself to variety and innovation in ways that bread did not. Unlike bread, cake was not a duty; it was a pleasure.

Housewives could express themselves with cake in ways they could not with bread. While adhering to a basic recipe, the baker can customize and personalize cake. She can use different flavorings, spices, and add-in ingredients: lemon and orange rind, raisins, cinnamon, ginger, nutmeg, mace, wine, and rum. One of the most commonly used flavorings until the middle of the nineteenth century was rosewater, a holdover from the medieval Arab influence on pastry.

Cupcakes, pancakes, waffles, fritters, and biscuits were individual foods, not communal. They fed Americans' desire for ownership of personal property. Unlike traditional breads, these breadstuffs were not broken from a common loaf and shared. Although biscuits that were chemically leavened might serve the same function in a meal as dinner rolls, they were vastly different in preparation. Yeast-risen rolls have to be scaled, so uniform pieces have to pulled off of the dough. To make sure they are identical, commercial bakers weigh them. Then, as the name indicates, the individual pieces have to be rolled out into balls. Commercial bakers roll out two at a time, one with each hand. A home baker has more leeway, but the rolls still need to be approximately the same size or they will bake unevenly; the small rolls will burn while the large ones will be underdone. Even for home bakers, making rolls requires an experienced eye and a practiced hand.

The dough for chemically leavened biscuits, however, can be flattened into a sheet and cut out with a biscuit cutter or the rim of a glass. This will make them identical. Or they can be dropped onto the baking sheet with a spoon. This will give them a more rustic appearance, but they will still be approximately the same size. With chemically leavened biscuits, technology, not the baker, does the work and creates consistency. Pans and spoons took the guesswork out of measuring.

However, all of these new leaveners came with caveats. They had serious drawbacks in acquisition, storage, and use. Both hartshorn and baker's ammonia create ammonia gas as a by-product. This can leave a urine-like taste in the food and an odor in the kitchen. Eliza Leslie advised against their use because they were "very unwholesome."[46] Their use now is limited to thin, crisp products such as European-style gingerbread cookies, because the small size allows the ammonia to escape and there are no dense fats to retain the odor. Because baker's ammonia is unpleasant, alternative leavening methods were preferred.[47] This formula is no longer used, although modern baker's ammonia has a similar ammonia content.[48]

By the second half of the nineteenth century, the use of emptins for leavening fell into disuse. Several factors were responsible. First, there was a decline in home brewing. German immigrants fleeing the failed revolutions of 1848 brought sophisticated brewing techniques for beers that tasted better than what Americans could brew at home. Germans established commercial breweries that produced affordable beer.

At the same time, the temperance movement in Jacksonian America cut back on the consumption of beer and other alcohols. In 1841 temperance devotee Sarah Josepha Hale commanded, "No *woman* should . . . [ever] make any preparation of which *alcohol* forms a part for family use!"[49] Lydia Maria Child (1844), on the other hand, continued the colonial tradition of home brewing and saw nothing wrong with adding homemade flavored brandy to her cakes, puddings, and pies, or rum to her pancakes. Child also believed that "beer is a good family drink."[50]

In the face of other, better leaveners, pearlash, too, began to fall into disuse. Beecher thought it the poorest of the chemical leaveners.[51] Eliza Leslie warned her readers: "All the alkalies, pearlash, soda, and sal-volatile, will remove acidity and increase lightness; but if too much is used, they will impart a disagreeable taste. It is useless to put lemon or orange juice into any mixture that is afterwards to have one of these alkalies, as they will entirely destroy the flavor of the fruit."[52] Catharine Beecher warned her readers too: "Always dissolve saleratus, or sal volatile, in hot water as milk does not perfectly dissolve it, and thus there will be yellow specks made."[53]

All of the alkalis also had cleaning and solvent properties. Pearlash could be used to clean painted woodwork, but "be careful not to scatter the pearlash water where it will lay any considerable length of time for it will dissolve or cut thro' the paint to the wood."[54] Child advised, "Sal-volatile, or hartshorn, will restore colors taken out by acid."

In the mid-nineteenth century, cream of tartar combined with baking soda seemed to be the solution to the leavening problem. This combination had the positive attribute of allowing the cook to use flavorings like lemon or orange. But it also had a serious negative: once cream of tartar came in contact with liquid, it immediately began to produce carbon dioxide. Any delay in getting the batter into the oven would cause the food to rise quickly then collapse. Beecher instructed: "bake *immediately*. . . . Try more than once, as you may fail at first."[55] This put further pressure on the baker.

Cream of tartar was susceptible to any liquid; even dampness in the air could render it useless. Beecher, a devout supporter of "sweet, well-raised, home-made *yeast bread*," also warned that badly mixed cream of tartar or an alkali could cause mouth or stomach problems, even though bread made with those was "good as an occasional resort."[56] Cream of tartar also had solvent properties. Child told readers to rub cream of tartar on white kid gloves to clean them.[57]

Chemical leaveners, like yeast, lacked consistency. Eliza Leslie repeatedly gave alternate amounts of pearlash in her recipes: "A small tea-spoonful of pearl-ash, or less if it is strong."[58] In her recipe for "Lafayette Gingerbread," Leslie advises: "Its lightness will be much improved by a small tea-spoonful of pearl-ash dissolved in a tea-spoonful of vinegar, and stirred lightly in at the last. . . . *—* If the pearl-ash is strong, half a tea-spoonful will be sufficient, or less even will do. It is better to stir the pearl-ash in, a little at a time, and you can tell by the taste of the mixture, when there is enough."[59]

The difficulties caused by inconsistencies in the leaveners were compounded by nonstandard measurements. Ingredients were measured in wine glasses, tea cups, pints, and pinches. Spoons were heaping, rounded, tea, soup, table, and coffee. All were further modified by size. Measurements varied not only across cookbooks but within them. For example, one of Leslie's recipes calls for "Half a large tea-cup full of best brewer's yeast," while another uses "A small tea-spoonful of pearl-ash."[60]

In spite of these difficulties, the new technology of chemical leavening shortcuts created new foods and new patterns of eating. Smaller, sweeter foods like cakes, cookies, pancakes, and soft quick breads proliferated. Catharine Beecher has several chapters devoted to breadstuffs: "Ovens, Yeast, Bread, and Biscuit"; "Breakfast and Tea Cakes"; "Plain Cakes"; and "Rich Cakes." Chemical leavenings also appear in the chapter on pie crusts and in the section titled "On Bread Making."

American women created a new type of uniquely American cake. Food historian Stephen Schmidt calls it "one of the most important cakes in American cake-baking history."[61] It was known by different names: "measure cake," "numbers cake," and "cup cake." It was a measure cake because the ingredients were measured, not weighed. It was a numbers cake because the recipe was easy to remember: "One cup of butter, two cups of sugar, three cups of flour, four eggs." It was a cup cake because it was measured in cups and baked in cups or small pans. It was easy to mix, and quick to cook: "Bake twenty minutes, and no more."[62] This was like a smaller version of pound cake, one of the few cake recipes in Child's *American Frugal Housewife* that did not contain a chemical leavener. In the twenty-first century, Child's recipe, with the addition of baking powder and milk, is still popular as the "1-2-3-4 Cake."

These new foods were eaten at new times and at different meals. In her novel *The Minister's Wooing*, about Calvinist society in Newport, Rhode Island, in the late eighteenth century, Harriet Beecher Stowe wrote: "Good society in New England in those days very generally took its breakfast at six, its dinner at twelve, and its tea at six." Tea parties began at three and lasted until sundown. Company tea began at 5:00 p.m., after the children were put to bed.[63] In the late eighteenth and early nineteenth centuries, Martha Ballard often served her baked goods when relatives, friends, or business associates came to tea. The diary contains 768 entries for tea and 47 for coffee. Chocolate appears a mere 12 times, often a gift "for former Services," or assistance.[64]

In antebellum America, however, Americans ate chemically leavened breadstuffs at breakfast, dinner, tea, and supper. "Cake was a principal attraction" at the new late-night parties that were now common social events.[65] Beecher devotes separate chapters to "Articles for Desserts and Evening Parties," and "Tea Parties and Evening Company," in which she instructs the hostess to place "cakes, pastry, jellies, and confectionary" on the table. She also says, "Set a long table in the dining-room, and cover it with a handsome damask cloth. . . . Set Champagne glasses with flowers at each corner. Set loaves of cake at regular distances."[66] Sarah Josepha Hale lamented that young women were ruining their health "by indulgence in these tempting but pernicious delicacies."[67]

Hale seemed to be of two minds. As bakeries proliferated, she advised women to take advantage of them: "Those ladies who live in the country must make their own cake; but for those who dwell in cities it is usually cheaper to buy it ready made for parties."[68] On the other hand, she also said that cake making, like bread making, should be "the province of the mistress of the house and her daughters."[69]

Foods made with chemical leaveners were eaten at new times, and they forced American women to think about time in a new way. Bread baking, like other artisanal work, was task-oriented. The process of bread making was slow, with breaks in between when the yeast rose. During the hours the yeast was working, the housewife could do other things: clean, garden, quilt. Hale advised needlework. A few minutes more or less in the rising time was not critical. With chemical leaveners, however, there was urgency. Once the leavener was in the batter, it had to go into the oven. Smaller breadstuffs had less margin for error in baking time too. Cake making, therefore, like the production of other goods in an industrial society, became time-oriented.

Chemical leaveners were a clear demarcation. The old, yeast-risen cakes were connected to bread and therefore to religion and the sacrament. The new cakes broke completely with the past. They had no history, no traditions or patterns of consumption, and most importantly, no connection to religion; they were completely secular. They were fun and were made for new fun social events like evening parties. The last vestiges of cake's ancient ancestral connection to bread linger only in nomenclature: gingerbread, quick breads.

Karen Hess, the doyenne of American culinary historians, objected to chemical leaveners. She was a purist and a traditionalist in general; she hated chemical leaveners specifically. Hess believed that in spite of chemical leaveners, there was nothing really new in American cooking. It was just a continuation of British cooking with new ingredients: "the only innovation being using an American product in place of a more familiar one"; the combination of "English culinary traditions" with "native American products."[70] She objected to the texture and the taste of chemical leaveners: "These chemical leaveners give lightness but an inferior texture to cakes.[71] She called these new cakes "a puffy debasement of older cakes."[72]

However, for Americans, altered texture was part of the allure of chemical leaveners. The new American chemically leavened cake achieved everything to which the British bread loaf merely aspired: it was white, light, and soft. Plus, it was sweet. Because it was light, it did not hold its shape and had to be baked in pans or molded tins. This ensured that it was pretty and never misshapen. Dark metal pans that retained heat also shortened baking time by as much as 20 percent.[73] Cakes could also be baked in a Dutch oven, a shallow cast iron pot with a lid. The batter could be put in the Dutch oven and then placed directly on the hearth. After the lid was put on the oven, the top was covered with coals, so the cake cooked top and bottom. The icing on the cake added a sweet white confection made from sugar and egg whites.

Hess also believed that chemical leaveners "leave faint metallic traces of bitterness—a taste that, unfortunately, Americans grew to love."[74] Perhaps Americans preferred the "faint" bitterness of chemical leaveners to the blatantly bitter bread and cake that hopped-up yeast and emptins produced. Hess believed that pearlash signaled the beginning of the end for American baking. Eighteenth- and nineteenth-century American cookbook writers disagreed with her. Eliza Leslie, in her recipes for yeast-risen breads and cakes, repeatedly instructed her readers to use pearlash, saleratus, or baking soda to correct sour yeast or dough that had risen badly. When Amelia Simmons included a valuable tip for bread makers—a way to rescue "grown flour" made from damp, damaged wheat, by adding pearlash—Hess commented, "It is, of course, not proper bread."[75] However, Hess commended British cookbook author Hannah Glasse for the "blessed lack of alkali [baking powder] in her baking, that noisome substance that came to dominate American baking."[76]

American women should be given more credit for what they created and for the chemical experiments they conducted in their kitchens. Even if pearlash was not revolutionary by itself—which it was—the accretion of innovation created a new American cuisine. Enough changes render the original unrecognizable and create something entirely new. This was the case with American cake and other breadstuffs. The lineage between a yeast-risen cake like Italy's *panettone* and America's current favorite, chocolate cake, is not apparent. Similarly, no one would say that chocolate mousse is not French because chocolate is not native to France, or that tomato sauce is not Italian because the tomato is not native to Italy and was not even in widespread use in Italy until the middle of the nineteenth century. A yeast-risen flat, round, English muffin has nothing in common with a sweet American muffin baked in a paper liner in a muffin tin. A fluffy American pancake leavened with baking powder and dripping with maple syrup is a far cry from its ancestor, a Normandy crêpe with stewed apples. Chemical leaveners are why the expression "flat as a pancake" has no meaning for Americans. A yeast-risen or spongy small cake is not the same as a chemically leavened cookie. American cookies are in a class by themselves.

The changes in leaveners were massive in the sixty-two years between 1796, when Amelia Simmons wrote *American Cookery*, and 1858, when the third edition of *Miss Beecher's Domestic Receipt Book* was published. Simmons used only three different types of leaveners—yeast, emptins, and pearlash; Beecher used eight. All three of Simmons's leaveners were produced by women in the home. Six of Beecher's eight leaveners had to be purchased outside the home (see appendix table A-2). Women had entered the market economy as innovators and producers. They quickly became consumers.

These autonomous American women, who had educated themselves and learned how to cook with corn and other strange flora and fauna, and who experimented with leaveners that could strip paint off floorboards, were not risk-averse. They were outspoken about their food, religion, education, and political issues. They were concerned with morality, healthy eating, and care of their families. Yet not one of these writers raised her voice in protest against pearlash, sal volatile, or any other chemical additive.

Why? Because chemical leaveners were a boon. They were convenient and inexpensive and were even regarded as healthier than yeast. Women were dissatisfied with chemical leaveners not because they were adulterants but because they did not work optimally. American housewife consumers did not want local, state, or federal governments to label, regulate, or remove chemical leaveners. They wanted *better* chemical leaveners.

CHAPTER 3

The Rise of Baking Powder Business

The Northeast, 1856–1876

The time required for a single person to prepare four loaves of a pound each, does not exceed five minutes, and the baking takes from thirty to forty-five more.

—Prof. Eben Horsford, Harvard University, 1861

The preparation of baking powder by Professor Horsford in Cambridge in North America, . . . I consider one of the most important and beneficial discoveries that has been made in the last decade.

—Baron Justus von Liebig, Germany, 1869

The slogan "Better living through chemistry" was created in the 1930s, but Americans had been living better through chemistry since at least a hundred years earlier, when chemistry was a new science.[1] On the eve of the Civil War, chemical leaveners were a commodity in search of an identity. Women had reached the limits of their amateur kitchen experiments with chemical leaveners. New and better chemical leaveners would have to come from the laboratories of men professionally trained in the new science of chemistry, and marketed in new ways by businessmen.

In post–Civil War America, an ideological shift from medieval dietetics and humoral theory to scientific nutrition propelled science, business, and food from the fourteenth century into the modern world.[2] As industry took over baking powder and other foods, the home shifted from the locus of production to the locus of consumption. Women continued to educate one another about the new chemical leaveners, using a new type of uniquely American cookbook that they created during the Civil War. By the time America celebrated its centennial, in 1876, three basic baking powder formulas had emerged, and baking powder had moved out of the drugstore and into the grocery store.

Science

> Most of [the American colleges that taught chemistry] were marching along daintily and grotesquely in the pointed shoes of the fourteenth century.
>
> —Prof. John W. Draper, President, American Chemical Society, "Science in America," 1876

In the early nineteenth century, everything about science was new, including the word "science" to mean the study of the natural and physical world.[3] In 1743, when Benjamin Franklin founded an organization to examine the natural world, he called it the American Philosophical Society; its members referred to themselves as "natural philosophers" in the Aristotelian sense.[4] A revolution in chemistry occurred in 1789 in France when Antoine-Laurent Lavoisier, the "Father of Modern Chemistry," published *Traité élémentaire de Chimie (*Elements of Chemistry), which laid the foundations for modern chemistry.[5] Lavoisier's investigation into human metabolism proved that the body's digestion of food was oxidation, which created heat and kept body temperature stable at 98.6 Fahrenheit.[6] Lavoisier also catalogued the elements, all 33 of them. (There are 115 now.)[7] Chemistry suffered a serious setback when Lavoisier was beheaded at the guillotine in 1794, a casualty of the revolution he supported.[8]

In spite of Lavoisier's efforts, chemistry remained the bastard child of the sciences. The other branches of science had theoretical and linguistic roots that reached back twenty-five hundred years to ancient Greece and classical antiquity. But the word "chemistry" was medieval and Arabic, related to "alchemy." Unlike physics and biology, which had produced beneficial machinery and medicines, chemistry was connected to poison and the dark arts.[9] In 1652 "chymistry" was defined as "a kind of praestigious, covetous, cheating magick."[10] The publication in 1818 of Mary Shelley's *Frankenstein*, about a chemist who brings a corpse to life, created the archetypal mad scientist as crazy chemist and exacerbated the idea of scientific experimentation as evil.

Americans had been suspicious of chemically adulterated food for more than half a century. In 1796, on the second page of *American Cookery*, Amelia Simmons warned readers about merchants who used chemical "deceits," such as painting the gills of market fish to give them "a freshness of appearance" in order to fool consumers.[11] In 1820, Frederick Accum, a British chemist, rang the adulteration alarm in England and the United States with his *Treatise on Adulterations of Food, and Culinary Poisons*. It was subtitled *Exhibiting the Fraudulent Sophistications of Bread, Beer, Wine, Spiritous Liquors, Tea, Coffee, Cream, Confectionery, Vinegar, Mustard, Pepper, Cheese, Olive Oil, Pickles, and Other Articles Employed in Domestic Economy and Methods of Detecting Them*. In the preface, Accum warned: "*THERE IS DEATH IN THE POT.*"[12]

The importance of bread in the diet is apparent from its first position in the title of Accum's book. Accum devoted several pages to bread adulterated with alum. British bakers used "sharp whites," flour that contained alum, so that they could "produce light, white, and porous bread, from a half spoiled material [flour]."[13] Accum described tests to determine if alum was present in bread and blamed adulterated bread for "the number of sudden deaths that are daily occurring."[14]

Chemistry was not regarded as useful to cooking either, even though all cooking is chemistry. Florence Nightingale found chemistry theoretical and irrelevant to nursing and no substitute for experience: "Chemistry has as yet afforded little insight into the dieting of sick. All that chemistry can tell us is the amount of 'carboniferous' or 'nitrogenous' elements discoverable in different dietetic articles. . . . But it by no means follows that we should learn in the laboratory any of the reparative processes going on in disease. . . . The patients [*sic*] stomach must be its own chemist."[15]

Even in academia, chemistry had no status in the early nineteenth century. American colleges taught little or no chemistry, so Americans who wanted to study chemistry went to France or Germany. Chemistry was not well regarded there either. German universities required an entrance exam for every subject except chemistry. Pharmacists used this back door to get into the university and benefit from the upward mobility afforded by a university education. It would take a master stroke to alter chemistry's image.

That came in 1840, with the publication of German chemist Justus von Liebig's book *Agricultural Chemistry*. Liebig's intention to elevate the reputation of chemistry and generate interest in it by connecting it to farming, the world's major occupation, was successful.[16] His discovery that nitrogen prevented soil depletion was at last a beneficial practical application for chemistry. It galvanized governments, universities, and private organizations, especially in the United States. In 1845 *Scientific American* magazine was founded. In 1846 Yale created its School of Applied Chemistry, which became the basis for the Sheffield Scientific School in 1854. In 1862 President Abraham Lincoln established the United States Department of Agriculture. The National Academy of Sciences followed a year later.[17]

Liebig was also one of the first chemists to examine the effects of food on physiology. In his 1848 book on nutrition, *Researches on the Chemistry of Food and the Motion of the Juices in the Animal Body*,[18] Liebig posited that a "dietetic trinity" of proteins, carbohydrates, and fats was sufficient to fuel the human body.[19] Therefore, the most highly valued foods were meat, breadstuffs and potatoes, and fats—a reflection of the northern European diet. Liebig's trinity remained gospel into the twentieth century, when vitamins were discovered and fruits and vegetables finally became valued in the diets of northern Europeans and Americans.

One of the young scientists who read Liebig's book on chemistry was Harvard professor Eben Horsford, an American and a former student of Liebig. Horsford

Harvard professor Eben Horsford, creator of Rumford Baking Powder.

was born on July 27, 1818, in Livonia (then Moscow), near Rochester, New York. He spent one year at Rensselaer Polytechnic Institute in Troy, New York, where he received a degree in civil engineering in 1838.[20] While teaching at the Albany Female Academy, he fell in love with a student, Mary L'Hommedieu Gardiner. But her father felt that Horsford's teaching job would not provide well enough for his daughter and withheld his consent to their marriage.

Horsford decided to get advanced, specialized training, so he went to Germany and studied with Liebig from 1844 to 1846, only the second American to do so.[21] Based on this cutting-edge education, even though he did not have a degree in chemistry, Horsford was offered the Rumford Chair for the Application of Science to the Useful Arts at Harvard in 1847. The benefactor of the Rumford Chair was Count Rumford (Benjamin Thompson), an American born in Woburn, Massachusetts, near Boston. A loyalist, Thompson had fled to Germany during the American Revolution and gained fame there for experiments with heat and friction, which resulted in the gas range and the pressure cooker.[22] At his death in 1814, Thompson left his papers to the United States government and his money to Harvard, which had allowed him to audit classes when he was a young man. The Harvard money was earmarked for a professorship "to teach the utility of the physical and mathematical sciences for the improvement of the arts and for the extension of the industry, prosperity, happiness, and well-being of society."[23]

Horsford's position at Harvard, with an annual salary of fifteen hundred dollars, removed the objections of Mary Gardiner's father to the marriage. On August 4, 1847, Horsford and Mary wed. By the time Mary died in 1856, they had four daughters. Two years later, Horsford married Mary's sister, Phoebe Gardiner, and they also had a daughter.[24]

Horsford's marriage was one reason that he was an outsider at Harvard: he had not married the daughter of a trustee or faculty member. He had also not attended Harvard as an undergraduate.[25] A third reason that Horsford was an outsider was that while philosophically Harvard preferred theoretical science, Horsford was keenly interested in the practical applications of chemistry.

Horsford patterned his laboratory after Liebig's; it was the first laboratory for analytical chemistry in the United States.[26] But he took his students out of the laboratory often: "He took his classes to visit local glassworks, soap factories, oil refineries, foundries, and gasworks 'where practical applications of the sciences to the arts may be advantageously witnessed.'"[27] Perhaps Horsford's position as an outsider at Harvard, combined with his own natural curiosity and inclination toward practical applications, made him amenable to a proposition from a Providence, Rhode Island, businessman. George Wilson believed that there was money to be made in chemicals and that Horsford should turn his knowledge of chemistry into marketable products.

Business

> [In the early nineteenth-century United States] business was carried on in much the same manner as it had been in fourteenth-century Venice or Florence.
>
> —Alfred D. Chandler Jr., *The Visible Hand*, 1977

Wilson, Duggan & Co. began in Providence, Rhode Island, in 1854 as a three-man partnership. The partners were George F. Wilson and J. B. Duggan. Horsford was the "& Co." In 1855 they established a plant in Pleasant Valley, Rhode Island, "for the manufacture of baking and medicinal products, chemicals used in dyeing and printing, and fertilizer."[28] Horsford continued to teach at Harvard while he went into business with Wilson.

It was no coincidence that this chemical factory began in New England. Geography and politics had conspired to make New England inhospitable to farming and hospitable to business. Glaciation had left the land rocky and uneven, difficult to farm but with waterfalls that could power mills. Samuel Slater opened the first textile mill in America in Pawtucket, Rhode Island, in 1793. In 1808, when the Embargo Act closed New England's ports, America's First Industrial Revolution began as entrepreneurs turned to manufacturing textiles.

Horsford was a new type of American, a "self-made man." When Henry Clay coined the term in 1832, it meant a man who had achieved his potential by his own efforts, "not necessarily in business and not just in monetary terms," although it came to mean that later.[29] Horsford was a specific type of self-made man, a new breed of scientist. The scientist was no longer an Enlightenment amateur gentleman engaged in intellectual pursuits in the privacy of his home using his own funds. Instead he was an academically trained professional who performed experiments to develop marketable products in an industrial laboratory financed by business. America's Second Industrial Revolution, in the second half of the nineteenth century, was based on the chemical creations of these new entrepreneurial scientists.

Horsford turned his attention to the bread problem. Like Sylvester Graham, Horsford knew that milling removed nutrients from flour. Unlike Graham, whose solution was to roll back time and use unsifted whole grain flour, Horsford sought to keep refined flour but use a modern chemical way to put nutrients back into breadstuffs. On April 22, 1856, after two years of experimentation, Horsford received a patent for the manufacture of monocalcium phosphate.[30] It fortified the flour. It was also a chemical leavener, to be used as "a substitute for cream-tartar or tartaric acid in the manufacture of yeast-powder or baking-powder."[31]

Horsford's baking powder was rich in calcium because it was made from beef and mutton bones that went through a lengthy process. First, bones were roasted and ground up in five-hundred-pound batches. The powder was boiled and stirred for three days until it became a paste-like mass. Then it was dried in the sun or artificially. After that it was ground into powder.[32]

Horsford's new powder differed from earlier chemical leaveners because it was made specifically to be a leavening agent and to fortify the flour. It was not simply a combination of preexisting ingredients that had other uses doing occasional duty as a leavener. He called it "yeast powder." It consisted of two ingredients, baking soda and the monocalcium phosphate made from bones. Like other chemicals, these ingredients were sold in pharmacies, where they were stored in bulk until the pharmacist measured out an exact amount and wrapped them in separate papers to prevent them from mixing prematurely. Paper is porous, which made the chemicals susceptible to moisture and diminished their potency. The packets were accompanied by a two-ended wooden measuring implement. The smaller measure was for the baking soda; the larger, for the phosphate. However, the ingredients still had to be measured separately, so there was room for error and confusion.

As Wilson, Duggan & Co. grew, the firm moved and changed structure. In 1858 the firm moved east to Seekonk, Massachusetts, which provided access to transportation on the Seekonk River. It was also in a more remote location, where fumes from processing bones at the plant would not annoy neighbors. On April 1, 1859, Wilson, Duggan & Co. disappeared, and the Rumford Chemical Works was incorporated "to carry on the manufacture and sale of drugs and chemicals."[33]

Rumford was capitalized at ten thousand dollars: one thousand shares at ten dollars each, fully subscribed.

Chemical leaveners received unexpected impetus from mid-century catastrophes in Ireland and France. The potato blight and ensuing famine in Ireland produced an atypical diaspora. When Europeans migrated, the men usually came first to in order to work and then later sent for their families. With the Irish, however, a disproportionate number of single women came to the United States. Unlike the women of other migrant groups, who remained in their own homes, the Irish women found work as maids and cooks in American homes.

Like other migrant groups, the Irish brought their food customs with them, including their bread. Irish soda bread was anomalous; all other European breads were leavened with yeast or sourdough starters.[34] Soda bread was a quick bread, leavened with baking soda. It was not baked in an oven, but cooked in a pan "in front of a peat fire, with the pot heated with peat underneath the pot and on the lid."[35]

As the father of five daughters, Horsford would have been aware that they spent a great deal of time baking. Perhaps they had household help. In Boston in the 1850s, chances were good that the help was Irish, a "Bridget" who had fled the potato famine and brought her knowledge of soda bread with her. Horsford was familiar with soda bread, but he did not like it, because "the proper adjustment of acid and soda is rarely if ever attained, and the bread is frequently bitter and portions of it brown, from the excess of alkali."[36] Isabella Beeton, the most influential British cookbook author of the mid-nineteenth century, who was also popular in the United States, added cream of tartar or tartaric acid to soda bread, which made Irish soda bread a baking powder bread.

SODA BREAD

1722. INGREDIENTS—To every 2 lbs. of flour allow 1 teaspoonful of tartaric acid, 1 teaspoonful of salt, 1 teaspoonful of carbonate of soda, 2 breakfast-cupfuls of cold milk.

Mode.—Let the tartaric acid and salt be reduced to the finest possible powder; then mix them well with the flour. Dissolve the soda in the milk, and pour it several times from one basin to another, before adding it to the flour. Work the whole quickly into a light dough, divide it into 2 loaves, and put them into a well-heated oven immediately, and bake for an hour. Sour milk or buttermilk may be used, but then a little less acid will be needed.

Time.—1 hour.[37]

The catastrophe in France that created interest in chemical leavening was sour wine. France's wine industry was on the verge of collapse. In 1859 Louis Pasteur was the first person to see under a microscope what had been present in wine making for thousands of years: yeast and bacteria. Pasteur's discovery of these two new organisms caused controversy in the scientific community. Pasteur believed these were the cause of wine fermentation and that they contributed to the rancid wine. Liebig disagreed. He considered yeast the "incidental concomitants of decay."[38]

Whether yeast was cause or effect, Horsford and other scientists agreed that "as a class, microscopic fungi are poisonous," and that not all of the yeast beasts were destroyed in bread baking. The public was revolted to discover that they had been ingesting live fungi. The residual living fungi would certainly run amok in the human body. Here at last was the cause of dyspepsia.[39]

The Theory and Art of Breadmaking: A New Process Without the Use of Ferment

If yeast was the cause, baking powder was the cure. Making bread with Horsford's yeast powder solved the dual problems of the bitterness of soda bread and of ingesting live fungi in yeast bread. In 1861 Horsford wrote *The Theory and Art of Breadmaking: A New Process Without the Use of Ferment,* a thirty-page treatise that explained his new process and its superiority. He began by explaining, with diagrams, that nutritious phosphates were stripped out of wheat along with the bran during the milling that made flour white. Dark breads like American graham, Westphalian pumpernickel, and Russian black bread, which contained bran, were the most nutritious, "in spite of their heaviness and sourness."[40] Horsford's powder would restore the nutritious phosphates to white bread without sacrificing lightness and sweetness.

Horsford then outlined the chronic difficulties of working with yeast and the history of the multiple options available to bakers for chemical leavening up to that time. He touted the advantages of his new yeast powder over all of them: consistency and convenience. Horsford saw that with his new powder, "breadmaking and most of the forms of pastry-preparation may be reduced to scientific precision."[41] He also saw that his powder would "make the preparation of all the forms of pastry a task of moderate skill and great certainty."[42]

The advantages of Horsford's new powder were the advantages of industrialization: minimal skill and time required to consistently produce a uniform product. In addition, bread made with yeast powder was better than yeast bread, because it was more nutritious, retained moisture longer, and could be eaten even by people with delicate digestions. It also allowed Americans to indulge in one of their favorite forms of consumption: "It may also be eaten warm with impunity," because it had never contained living organisms.[43]

Horsford illustrated his point with two recipes. The first is for the French method of making yeast bread. It requires multiple steps of measuring, resting, and kneading over two days. Then, five successive batches of bread must be baked to arrive at "fancy"—that is, white—bread. The second recipe is what Horsford calls the "New Method." It takes less than one hour from mixing to consuming, in only three easy steps.

HORSFORD'S NEW METHOD [OF BREAD MAKING]

1. Provide a quick oven. [This requires a temperature of from 350 to 450 Fahr.]
2. Stir a measure each of acid and soda into a quart of sifted dry flour, to which a teaspoonful of salt has been added. Mix intimately with the hands. Then add from time to time cold water from a pint cup, stirring and kneading meanwhile, until just a pint of water has been most thoroughly incorporated with the flour.
3. Shape the mass of dough into a loaf, place it in a deep tin bake-pan, with a cover so high as to be out of the reach of the risen dough, and set it immediately in the oven.

This new bread could not be "cast"—put directly on the oven floor, like yeast bread—but had to be put in a pan. Yeast bread shaped into a round loaf with the hands retains its shape because the gluten binds the flour as it rises. Chemically leavened bread, which does not go through a lengthy rising process, is amorphous and needs a pan to contain it. The need for pans for baking powder breads and quick breads contributed to the tin industry and the manufacture of new styles and shapes of pans.

Horsford also sold "Bread Preparation," which he considered a self-rising flour. This was eleven ounces of his acid and baking soda, which the baker was to mix into twenty-five pounds of flour. When the baker needed flour, it was ready, with the leavener in it.[44]

In 1862 the Rumford Chemical Works changed its address but not its location to East Providence, Rhode Island. The Massachusetts–Rhode Island boundary had shifted; the firm had to reincorporate in Rhode Island. The following year, the business was profitable enough that Horsford was able to resign the Rumford Chair at Harvard and devote his energies to the chemical works full-time.[45] On November 6, 1865, the company became an equal partnership between Horsford and Wilson.[46]

The Civil War interrupted Horsford's experiments with baking powder as he turned his attention to trying to find ways to feed the troops. After the war ended in 1865, he returned to baking powder. In 1869 he found a solution to the problem of two separate ingredients: he added a third, neutral ingredient, cornstarch.

Cornstarch, like baking powder, was an American creation. Its use in baking powder was not an obvious choice. Although cornstarch is commonplace now, in 1869 it was a new product, the result of cutting-edge chemistry. Corn technology was in its infancy. Chemists were just beginning to deconstruct kernel and cob into the great maize triumvirate: corn starch, corn syrup, corn oil. The name "cornstarch" is not a description of the product, which is a soft, white, tasteless powder, but of its original use by commercial laundries to stiffen fabric. In 1842, Thomas Kingsford, a British immigrant who worked in a wheat starch extraction factory in New Jersey, applied the same technology to extract starch from corn. In 1848 he opened his own cornstarch factory in Oswego, New York, on Lake Ontario, which had "pure water, excellent shipping facilities and water power."[47] From there, cornstarch manufacture spread throughout the Midwest, but it was not until approximately 1850 that Kingsford modified cornstarch to make it a food-grade product.[48]

Horsford's addition of cornstarch was revolutionary in several ways. It made baking powder stable because it acted as a buffer to prevent the two leavening agents, baking soda and phosphate, from combining and producing a chemical reaction—exploding.[49] The cornstarch also absorbed moisture, which extended baking powder's shelf life and kept its potency longer. This stability allowed the ingredients to be premixed into a single powder that could be sold in one container. Horsford further extended his product's shelf life by placing it in a glass bottle with a stopper, which provided more protection from moisture than the paper wrappers in which chemicals were usually sold.[50] Also gone was the two-ended measuring spoon and the confusion. One spoon would suffice.

This was the beginning of the shift from bulk to individual packaging that is standard today because it has benefits for both businesses and consumers. Unlike generic chemicals scooped out of bins by pharmacists, these individual containers could be labeled with a unique brand. Individualization also created demands for bottles and other packaging, label design, printing, and eventually, advertising. For consumers, a single powder in a bottle provided increased sanitation, ease of storage, and ease of use. A label with a company name, in addition to being a brand for the business, removed anonymity and provided security and accountability for the consumer.

Horsford's addition of cornstarch meant that women no longer had to make a special trip to the chemist to buy separate leavening ingredients. It made more sense to sell baking powder where flour and sugar were sold. Horsford's innovation moved baking powder from drugstore to grocery store, from pharmaceutical to food. Here at last was the easy-to-use, convenient, consistent chemical leavener that housewives had been seeking.[51]

By 1869 Rumford was doing better but was still "scarcely afloat." The company produced only 1.5 million pounds of baking powder.[52] This was in spite of sales in England under the name "Horsford-Liebig Baking Powder." Liebig was so well known internationally that there was skepticism regarding the product. Horsford was viewed as an interloper trying to steal credit for Liebig's creation. In 1869 a London journal reported: "The prefix, of 'Horsford' is used in conjunction with that of Liebig, Horsford claiming to be the original inventor of that which Liebig has improved." To reassure the consumer, the writer added: "It is understood that the article is prepared in Germany, under the supervision of Liebig himself."[53]

Modern Rumford Baking Powder cans still sport the nineteenth-century silhouette of Count Rumford (Benjamin Thompson).

This skepticism and the attendant controversy over who invented baking powder remained until 1993, when Professor Paul Jones wrote the definitive essay on the relationship between Horsford and Liebig. Jones used correspondence between Horsford and Liebig as well as Liebig's own words to prove that Horsford was the inventor: "The preparation of baking powder by Professor Horsford in Cambridge in North America, . . . I consider one of the most important and beneficial discoveries that has been made in the last decade."[54]

Horsford revered Liebig and in 1869 provided him with the formula for baking powder in the hope that Liebig could introduce it in Germany and profit from it financially.[55] Attempts to market baking powder in Germany are a lesson in American exceptionalism. The United States was a new country, eager to find a national identity and to differentiate itself from its European roots. Germany, like the other European countries, had centuries of traditions, including food traditions. Unlike the United States, where housewives were the main market, German baking powder companies had to contend with male bakers' guilds entrenched for centuries, with disincentives to change. It was not simply a matter of technology. To embrace a new method of bread making, the German people would first have to be convinced that there was something wrong with the bread they had been consuming every day for generations, and the bakers would have to admit that they did not know how to bake bread. Neither of those was going to happen.

German businessmen began to lose interest in baking powder when they realized that the only way they could convince professional bakers to overcome their prejudice against it would be to "travel from city to city over one to two years, showing bakers individually how to work with baking powder and revealing to them what advantages it would provide."[56] These methods of introducing consumers to new goods and educating them were viewed as opportunities and standard methods of doing business in the United States, where the Yankee peddler and the peripatetic patent medicine salesman, or "drummer," were well-known figures. However, in Christian Europe, opprobrium was connected to selling goods door-to-door: itinerant peddlers were Jews.[57]

Horsford and Liebig had also hoped that physicians would endorse baking powder bread as more nutritious and easier to digest than yeast bread, but the German medical profession saw nothing wrong with German bread and had no interest in baking powder either. Therefore, manufacture of baking powder in Germany remained small. In 1869 German baking powder companies sold approximately thirty-two thousand pounds. By 1872, sales had dropped to just twenty-seven hundred pounds, most of it exported because "only one baker in Germany was still buying."[58] With no economies of scale or scope, baking powder was more expensive than yeast and did not find a market in Germany at that time.

Beginning in 1869, Rumford also had expenses for patent infringement lawsuits the company initiated in New York, South Carolina, Georgia, New Jersey, North Carolina, Maryland, Tennessee, Rhode Island, Massachusetts, and Minnesota. Horsford discovered these imitations through the painstaking experiments he did almost daily to analyze the chemical composition of his competitors' products. He also constantly sought to improve his own formula and experimented with different types and amounts of starch and baking soda, and even cream of tartar. He sometimes did as many as five experiments in one day.[59]

By the end of the 1870s, Rumford was on much more solid financial ground. In 1876 the company began "active exertions" to expand its sales to Chicago, St. Louis, Memphis, and New Orleans.[60] At the board of directors meeting on July 20, 1877, the stock was worth $340,000: 3,400 shares at $100 each.[61] Although Rumford's business was growing, the cream of tartar and baking soda combination that housewives had stumbled upon decades earlier still had the majority of the market.

Royal: The Cream of Tartar Formula

> The mixture of dry tartaric acid and bicarbonate of soda with flour is well known to produce a light and palatable bread.
>
> —Eben Horsford, *Bread-Making*, 1861

With knowledge widespread about the chemical leavening reaction that occurred when cream of tartar or tartaric acid and baking soda were combined, anyone who understood the principle could create baking powder. For this reason, cream of tartar–based baking powders were common in the industry's early phase. The name "cream of tartar" derived from the process by which it was created:

> During the fermentation, in the process of wine making, an acid material settles, in the form of a crust or scales, on the sides and bottoms of the casks. It is this material, known as argols, from which cream of tartar and tartaric acid are made. The

> argols are carefully washed, dissolved in boiling water and drawn off into copper tanks. As the solution cools, a thick crust forms over the entire surface, which is taken off like cream from the surface of milk; hence, the name, "cream" of tartar. It is then re-dissolved, purified, filtered, again crystallized, powdered, and sold under the name of "cream of tartar."[62]

In 1866, in Fort Wayne, Indiana, druggist Mr. Thomas M. Biddle and businessman Joseph C. Hoagland created the partnership of Biddle & Hoagland to make a cream of tartar baking powder in their drugstore. A few months later, Hoagland's brother Cornelius joined the business. Biddle and Cornelius made the baking powder by hand in five-pound batches, using a mortar, and sold it over the counter by the ounce or the pound. The following year, the company split. The Hoaglands sold the drug business to Biddle, kept the baking powder business, and moved to Chicago. They began to use the name "Royal Baking Powder" and sold it in Illinois, Michigan, Indiana, and Ohio. In 1868, with their sights set on a larger market, they moved to New York City and created the Royal Baking Powder Company at 71 William Street. The company expanded with the hire of two sales representatives to cover New York and Brooklyn.[63]

One of the salesmen was William Ziegler, the son of German immigrants. Ziegler was born in 1843 in Beaver County, Pennsylvania, near the Ohio border. The family moved to a farm near Muscatine, Iowa, and his father died when Ziegler

William Ziegler, Royal Baking Powder.

was three. When Ziegler was thirteen, he worked as a printer's apprentice in a newspaper office, then later as a druggist's clerk. He moved to New York City in approximately 1863 and attended either the New York School of Pharmacy or Eastman's Business College in Poughkeepsie.[64] These ventures trained him well for the new economy and for the baking powder industry. In 1868 Ziegler partnered with John H. Seal to start a bakery supply business. Shortly after, Royal hired them as salesmen working on commission.

Ziegler and Seal worked for the Royal Baking Powder Company for a few months. Then they began to sell baking powder under the name Royal Chemical Company. When the Hoaglands sued Ziegler for using the Royal name, Ziegler testified that he and Seal had "manufactured baking powder prior to selling that of the Royal Baking Powder; we adopted the name before we sold their goods at all, before I knew the Hoaglands at all."[65] In March 1873 the suit was resolved by merging the two companies under the name Royal Baking Powder Company. The company was capitalized with sixteen hundred shares of stock at a par value of one hundred dollars each.[66] They set up offices and a small factory at 178 Duane Street in lower Manhattan. Cornelius Hoagland was president, Joseph Hoagland was treasurer, Ziegler was secretary, and Seal was in charge of the factory. Royal Baking Powder prospered, and the company grossed $385,162.18 in just ten months in 1873, the year the United States began to celebrate its centennial.[67]

Female Producers and Consumers

> Baking Powder—Eight ounces flour, eight of soda, seven of tartaric acid; mix thoroughly by passing several times through a sieve.
>
> —Mrs. Trimble, Mt. Gilead

> Baking Powder—Cream of tartar two parts, bicarbonate of soda one part, corn starch one part, mix.
>
> —Mrs. B. H. Gilbert, Minneapolis

Women launched the centennial celebration in Philadelphia on December 17 and 18, 1873, with a tea party to commemorate the Boston "tea riot" of December 16, 1773. Male speakers praised "the peculiar excellence of American inventions, such as the steamboat, the telegraph, the cotton gin, the rotary printing press, the carding machine, the grain reaper." They should have included baking powder, because without it there would have been few refreshments at the national tea party. Served with the tea were "biscuit, sandwiches, crullers, cake, ice cream, etc." Most was donated. The tea cups were white, with a "*fac simile* [*sic*] of the signature of John Hancock," and were sold for twenty-five and fifty cents to raise money for the fair. There was a miniature replica of the ship the *Dartmouth*, with tiny Indians throwing tea chests overboard. The District of Columbia display was a giant cake "surmounted by an image of the Goddess of Liberty, and supported by the figures of Washington and Franklin."[68]

To celebrate the centennial and pride in their American identity, including their culinary identity, Americans resurrected the British upper-class tea and refashioned it as their own. When Maine resident Martha Ballard wrote in her diary in the early 1800s that people came for tea, she meant the working-class British light evening meal of supper, often simply a modest slice of pie. In the Gilded Age, however, Martha Washington balls and tea parties were occasions for conspicuous consumption.[69] The balls were held on or as close as possible to George Washington's birthday, February 22. They were elaborate evening affairs, with the costumed host and hostess impersonating George and Martha Washington. Attendees were masked, wigged, and powdered, more Versailles than Mt. Vernon. Participants recited poetry that reinforced American ideology and mythology about colonial roots, the Victorian home, and Washington, honesty, and the cherry tree. They also consumed baked goods, especially cake made in the new American style, leavened with baking powder.[70]

In contrast to the balls, Martha Washington tea parties were women-only events, with the hostess assuming the role and guise of Martha. The teas were more informal afternoon affairs that replaced five o'clock tea. Etiquette manuals decreed that they could be held "any time during the season for balls, fairs, and general festivities." It was also acceptable to use the tea parties as fund-raising events.[71] Martha Washington tea parties continue into the twenty-first century. In addition to ongoing historic reenactments of Tea with Lady Washington at Mt. Vernon, Virginia, Martha Washington tea parties are also the theme for bridal showers and parties for the Daughters of the American Revolution in Connecticut, and debutante balls in Laredo, Texas.[72]

The increase in baking powders of all types was reflected in the *National Cookery Book*, published in 1876 for the centennial celebration in Philadelphia. The *National Cookery Book* is an example of a new type of uniquely American cookbook that American women created as a grassroots movement during the Civil War. Community cookbooks are cooperative efforts by women to raise money for charity. These cookbooks were previously called compilation cookbooks because they are not written by one woman, but compiled by an editor from recipes contributed by many women, credited or uncredited. The first known cookbook of this type, Maria J. Moss's *A Poetical Cook-Book*, was published in 1864 to raise money for the Philadelphia Sanitary Fair.

These cookbooks are also commodities, products created by women to raise money. Like the New England town meeting that Thomas Jefferson believed was the perfect example of direct democracy, community cookbooks are perfect examples of the American combination of democracy, capitalism, and Protestantism.

Community cookbooks brought women into the public sphere and allowed them to achieve a degree of recognition that most men never achieve in their lives: their names in print. In the twenty-first century, "hundreds to thousands of such books were being published annually."[73]

One of the managers of the Philadelphia Sanitary Fair during the Civil War was the motive force behind the *National Cookery Book*. Elizabeth Duane Gillespie, the great-granddaughter of Benjamin Franklin, compiled the recipes, winnowing them down to about 950 from thousands contributed by women from every state and the territories.[74] The purpose of the *National Cookery Book*, which was dedicated to "The Women of America," was to explain American cuisine to foreigners who were curious about what Americans ate.

Although there were seemingly seven different types of leaveners in this cookbook, the single most common ingredient was baking soda, called both soda and saleratus. It was in all of the non-yeast leaveners. The antebellum cream of tartar and baking soda combination that a housewife could make in her kitchen was still popular, with fourteen recipes. However, the new industrial premixed leaveners, called baking powder and yeast powder, lagged by just one recipe. Total non-yeast leaveners outnumbered yeast by almost two to one. Yeast was used primarily for bread. In cakes and muffins, however, non-yeast leaveners predominated (see appendix table A-3).[75]

How women viewed the new baking powders is reflected in the names of the recipes. One is for "Feather Cake," as in "light as a feather." Another recipe is for "Hocus Pocus Pound Cake," a twelve-egg cake made with one pound of sifted flour: "The hocus pocus is one teaspoonful of azumea or baking powder, sprinkled over the batter and well beaten in."[76] To this woman, baking powder was a magic potion. And she knew and recommended it by its local brand name, Azumea.

The *Centennial Buckeye Cook Book*, a community cookbook published in 1876, began modestly as a fund-raiser to complete construction of the First Congregational Church in Marysville, a small town in Ohio. The recipes were contributed by the church women. However, in the next quarter century, the *Centennial Buckeye Cook Book* became "the largest-selling American cookbook in the nineteenth century"—more than one million copies in nine revisions.[77]

What accounted for this cookbook's astounding popularity? As Andrew Smith has pointed out, it was aggressively marketed in newspapers and at centennial celebrations even before it was printed. In modern marketing terms, the book had prepublication buzz. Smith also pointed out that the book is more than four hundred pages long and comprehensive. This was in contrast to most charitable cookbooks, which were often only fifty to one hundred pages and contained only recipes, not cooking instructions.[78]

Some of the recipes reflect the influence of Connecticut and Massachusetts in the Western Reserve, the Midwestern territory that eastern colonial states had claimed as far west as the Mississippi. There are fourteen recipes for graham, brown, and corn breads. There is a recipe for soda bread, called "Poor Man's Bread."[79] The recipe for "Old Hartford Election Cake (100 years old)" was copied from Catharine Beecher and was mislabeled. The recipe was already one hundred years old when Catharine Beecher included it in her cookbook decades earlier, shortly after she left Ohio.[80] But gone is the distinction that Beecher had made between rich and plain cakes; in *Buckeye* they are all just cakes.

However, many of the recipes in *Buckeye* were new, "on the cutting edge of culinary life in America."[81] Baking powder was an important factor in what made them new. The *Centennial Buckeye Cook Book* shows that by 1876, the transition from yeast to baking powder was complete in the Midwest. More than half of the recipes for bread—45 out of 75—use chemical leaveners. The cakes, including variations on European pound, sponge, wedding, and fruit cakes, have become distinctly American, almost universally leavened with either baking powder or baking soda. Baking powder by name was the single largest category of leavener for cakes and other non-bread foods, with 46 recipes. If yeast powder, and cream of tartar plus soda, which equals baking powder, are added, 80 of the 139 recipes use baking powder (see appendix table A-4). From the ratios of baking powder to flour, it is clear that most of the baking powder these women were using was not cream of tartar–based, like Royal's. In addition to baking powder, cakes are further lightened by replacing some of the flour with the new American ingredient, cornstarch.

Several new cakes are included in this cookbook. Marble cakes are made of two different contrasting batters, combinations of white and dark, white and chocolate, or gold and silver swirled together attractively. There are new layer cakes, some as high as five layers, with custard filling between the layers. "Snow Cake" is another popular new cake, the forerunner of what we now call angel food cake. Jelly roll cakes—sponge cakes spread with jelly or other sweet filling and then rolled up—also appear in this cookbook.

Other reasons the cookbook sold so well were demographics and American exceptionalism. In 1870 the mean center of population in the United States was in western Ohio, near Marysville, where the book was written. By 1890 the mean had moved to Indiana, where it remained until 1950.[82] Also, literacy was highly valued in the United States. Unlike European towns, which had been built around churches, with surrounding open space that served as a marketplace, American towns in the Old Northwest Territory—Ohio, Indiana, Illinois, Michigan, and Wisconsin—were surveyed, planned, and built around schools. Also unlike Europe, where state religions were supported by state funds, American constitutional separation of church and state meant that churches had to be financially self-sufficient. Church women discovered that community cookbooks were a good way to raise money.

By the time America celebrated its centennial in Philadelphia in 1876, two basic formulas for commercial leaveners had emerged: (1) Rumford's phosphate, from burned bones, and (2) Royal's cream of tartar, a by-product of wine production. They both also contained baking soda and a cornstarch buffer. Scores of companies scattered throughout the United States manufactured these leaveners in various permutations with different potencies and also experimented with other chemical formulas. These were not interchangeable. Nomenclature also contributed to the confusion because they were called "baking powder," "yeast powder," "yeast substitute," and "bread powder."

Competition existed not only among baking powder companies but also between baking powder companies and women. Housewives presented a two-pronged problem for baking powder companies. Women who were skeptical of chemicals, as well as self-sufficient women who mixed their own baking powder, were obstacles to market expansion. Baking powder companies would have to educate these women to be not producers, but consumers. To pry these antimodernist women loose from their old ideologies, baking powder companies would have to convince the skeptics that baking powder was safe and efficacious, and convince the self-sufficient bakers that their homemade baking powder was inferior to what professional chemists in a laboratory produced. Baking powder companies would have to wage war on two fronts: against each other, and against recalcitrant female consumers.

The one weapon in the two wars would be advertising.

CHAPTER 4

The Advertising War Begins

"Is the Bread That We Eat Poisoned?" 1876–1888

The Baking Powder War: Is the Bread That We Eat Poisoned by Alum or Is It Not?—When Doctors Disagree, Who Shall Decide? . . . A Question of Interest to Every Family

—*[Brooklyn] Daily Eagle*, 1878

The pioneers . . . in advertising, were the patent medicine men, with circuses and tobacco vendors close behind.

—James Harvey Young, *The Toadstool Millionaires*, 1961

During the Gilded Age, a time of increasing secularity, the American dream began to be measured in merchandise. On the surface, advertisers were selling baking powder, patent medicines, or cereal. However, they were redefining "the American dream in terms of a consumption ethic."[1] In the case of baking powder, it was a literal consumption ethic. Baking powder's role in this ethic was to provide the means to a life of the luxurious baked goods that previously had been available only at great expense of either time or money. With baking powder—the *right* baking powder—all luxury and indulgence were achievable and affordable; life could be a party. Cake was one symbol of a rise in the standard of living that occurred in Gilded Age America, which saw a dramatic increase in the buying power of the food dollar: "in 1898 one dollar could buy 43 percent more rice than in 1872, 35 percent more beans, 49 percent more tea, 51 percent more roasted coffee, 114 percent more sugar, 62 percent more mutton, 25 percent more fresh pork, 60 percent more lard and butter, and 42 percent more milk."[2]

Advertising was the ideological warfare that businesses waged to get more of this consumer dollar. Truth in advertising was not a concept in the nineteenth century. There was no government regulation of advertising claims, just as there was no regulation of food or drugs. Patent medicine salesmen made fantastic, hyperbolic claims. Their products could cure headaches, weight loss, back pain, dyspepsia, nervousness, and cancer, and they had the fake statistics to prove it.[3] Patent medicines and soothing syrups did make people feel good in the short run, because they contained as much as 23 percent cocaine, morphine, and/or alcohol. Take four times a day.[4]

As with science and business, nineteenth-century advertising straddled the premodern and the modern. Baking powder companies, like other American businesses, used the advertising tactics that had been introduced by peripatetic patent medicine salesmen. With post–Civil War urbanization, advertising became more sophisticated as businesses sought new ways to brand their products and woo consumers. Men at baking powder companies countered women's community cookbooks and pioneered a new form of advertising: the corporate cookbook. New technologies such as color lithography made pictures available in addition to text on another new form of advertising: trade cards.

Before newspaper advertising became widespread after the Civil War, businesses reached consumers through personal contact and tangible giveaways. They used "joke-books, cook-books, coloring books, song-books, and dream-books . . . handbill ballads . . . pill-filled paperweights . . . decorated porcelain . . . a china platter."[5] The Royal Baking Powder Company also relied on "canvassing from house to house; personal solicitation in every way." Royal salesmen would "paint signs on brick walls, deliver samples to houses, . . . bill posters, decorate grocery stores, and every way that suggested itself; [they] delivered samples; [they] would put the baking powder up in small packages and deliver it from house to house for a sample to be tried."[6] They also used gutter snipes—that is, posters affixed to curbs—so that pedestrians would see them as they waited at street corners or crossed the street.

In spite of these blitz attack sales efforts, by 1876 Royal's profits dropped and the company faced a crisis. There was a powerful new weapon in the baking powder war: a chemical called alum. Cream of tartar cost thirty cents per pound; alum cost three cents.[7] New companies using the new formula proliferated and challenged Royal's lead in the market: "Of late the cream of tartar baking powder companies have been surprised by the introduction of a number of other baking powders, sold by small but rising baking powder companies at cheaper rates."[8] By 1878 there were at least forty-two baking powder companies in the New York City area alone.[9]

Royal went on the offensive and began an advertising war in the press.[10] The pioneering advertising campaign played on the public's fears about chemicals in general and adulterated food in particular, which had been present for almost a century. Although the argol in Royal's cream of tartar was a chemical sold in a pharmacy and Royal had started in a pharmacy, Royal took great pains to distance itself from chemicals in its advertising. It presented Royal as "from the grape" and pure. Therefore, Royal was free to point its finger and sound the chemical warning bell about its competitors' products with impunity. The press was already filled with horror stories about pharmaceutical accidents and wrongdoings that resulted in death. Beginning with Rhode Island in 1870, states passed laws to regulate drugs and pharmacists. Lawmakers were especially concerned with the adulteration of drugs, and with poisons, including poisons in food.[11]

In this atmosphere of fear about chemicals, Royal planted anti-alum articles in the press. A headline in the Brooklyn *Daily Eagle* proclaimed, "The Baking Powder War: Is the Bread That We Eat Poisoned by Alum or Is It Not?—When Doctors Disagree, Who Shall Decide?"[12] The article contained interviews with scientists who said that "alum" was a "cheap substitute" for cream of tartar and was injurious. The article was reprinted word for word in other newspapers.[13]

Both William Ziegler and Joseph Hoagland took credit for initiating the baking powder war. Ziegler said that Hoagland had objected to it because he felt that it would reflect badly on all baking powders and damage the market. Hoagland denied this and provided more details about the campaign to back up his claim, including that Royal had conducted alum experiments on dogs. They got ten dogs from the pound and fed them for three weeks on baking powder biscuits made with alum. At the end of that time, the dogs were sick and the scientists Royal hired published their findings that alum baking powder was the cause.[14]

In addition to newspaper advertising, in 1877 Royal published *The Royal Baker and Pastry Cook*. The cookbook was a declaration of war against all other baking powders. It was also a new kind of uniquely American cookbook: the corporate cookbook. William Ziegler was an advertising genius who created a masterpiece of marketing in the form of a cooking pamphlet. *The Royal Baker and Pastry Cook* used a multitude of techniques to promote Royal's products and discredit the competition in 40 pages and 377 recipes. Through aspirational appeal, testimonials, and financial comparisons, it allayed consumer qualms about chemical leaveners and "proved" that Royal was the best baking powder.[15]

In 1876 the *National Cookery Book* stated what every woman knew was the truth about baking: "experience must of course prevail over teaching."[16] The following year, *The Royal Baker and Pastry Cook* trumpeted the opposite: with baking powder, no experience was necessary. Baking powder conferred instant expertise on everyone equally. This American invention was a truly democratic product.

The Royal Baker and Pastry Cook was a longer, more sophisticated version of eighteenth-century agricultural almanacs and patent medicine cookbooklets. The recipes in those publications had nothing to do with the almanac or the medicine; including them was an advertising ploy to get the attention of housewives and to prevent the almanac from being thrown away. Almanacs began printing recipes as early as 1795; patent medicines by the middle of the nineteenth century.[17] By the last quarter of the nineteenth century, manufacturers of food and food equipment used the recipes in cookbooklets to advertise and educate customers about their products: "The patent medicine almanac was a sort of informal textbook for educating the American people." From ten to thirteen million copies of almanacs alone were printed in the Gilded Age.[18]

The Royal Baker and Pastry Cook also cross-promoted Royal's other business: selling extracts of herbs, spices, fruit, and flowers. The herbs were powdered; the spice extracts were liquid and came in lemon, peach, ginger, celery, vanilla, orange, nutmeg, almond, rose, nectarine, cinnamon, and clove. From the large quantities necessary in each recipe, they were probably not cost-efficient. Storage would have been a problem, too, before screw-top bottles. Bottles with stoppers would have allowed evaporation and dissipated the strength of the extract.

The Royal Baker and Pastry Cook was directed at educated middle-class women in several ways. First, it weeded out the riffraff up front by costing ten cents instead of being free. Then it addressed the chronic servant problem by promising that Royal baking powder would provide a solution to the problem of "servants [who] are not particular" when they measure cream of tartar and baking soda.[19] It also included one page of advertising for the latest cooking equipment. These are the only images in the book; even the cover is plain blue paper and contains only text. The cookbook itself is text-dense. The cake pans, muffin rings, and molds are for professional-style cooking in the home. They are expensive; the timbale mold costs two dollars. An ice cream freezer costs five dollars, and a "family scale" is four dollars.[20] The scale would not be needed for any of the recipes in this booklet, however, because they are standardized in cups and spoons; no longer were recipes by weight. Americans were mobile; a scale was cumbersome to transport on a wagon. But everybody had a cup and a spoon.

The Royal Baker and Pastry Cook was patterned after the most popular cookbook in England at the time, and "the most famous English domestic manual ever published," *Mrs. Beeton's Book of Household Management*. First published in 1859, almost two million copies had been sold by 1868.[21] All of the recipes in Isabella Beeton's cookbook were numbered; so are the recipes in *The Royal Baker*. *Mrs. Beeton's* listed exact measurements at the beginning of each recipe, followed by explicit instructions; so does *The Royal Baker*. The British format was just part of *The Royal Baker and Pastry Cook*'s continental allure. Many of the foods have foreign influences and names: "Vienna Rolls," "French Rolls," "London Crumpets," and "Croûtes en Diable (Deviled Toast)," among others.

At the same time that it appealed to traditional European tastes, *The Royal Baker and Pastry Cook* also fed Americans' desire for novelty. The cover promised: "The recipes in this book are new." They were new in the sense that they all use Royal baking powder. Even "Election Cake," traditionally made with yeast, had been reworked for baking powder. So had recipes for pie crusts. There is only one cake

THE ROYAL BAKER AND PASTRY COOK.

The cuts on this page represent pans used in the various kinds of baking, and are referred to in the Receipts according to numbers.

Fig. I. 2 QT. CAKE MOULD. Price..........50 cents.

Fig. V. TIMBALE MOULD. Price..............$2.00

Fig. IX. LEMON CAKE PAN. Price, doz..........$1.00

Fig. XIII. SQUARE CAKE PAN. Price..........50 cents.

Fig. XVII. MUFFIN RINGS. Price, doz.....50 cents.

Fig. II. 3 QT. PUDDING MOULD. Price...............$1.25

Fig. VI. PUDDING MOULD. Price...............$1.00

Fig. X. CAST GEM PANS. Price..........50 cents.

Fig. XIV. BAKING SHEET. Price..........50 cents.

Fig. XVIII. OVAL TIN PAN. Price..........30 cents.

Fig. III. 2 QT. CAKE MOULD. Price..........50 cents.

Fig. VII. MUFFIN PANS. Price...............$1.00

Fig. XI. WASH BRUSH. Price..........75 cents.

Fig. XV. ONE GALLON ICE CREAM FREEZER. Price...............$5.00

Fig. XIX. WAFFLE IRON. Price..........75 cents.

Fig. IV. 3 QT. CAKE MOULD. Price..........50 cents.

Fig. VIII. OVAL PUDDING PAN. Price..........30 cents.

Fig. XII. TIN BREAD PAN. Price..........30 cents.

Fig. XVI. FLOUR SIEVE. Price..........50 cents.

Fig. XX. FAMILY SCALE. Price...............$4.00

Any of above can be purchased at tinware or house furnishing establishments; and failing to get them at above places, we will send them by express, securely packed, on receipt of price.
In ordering from us, give figure number only.

Address,
ROYAL BAKING POWDER COMPANY,
No. 171 Duane Street,
P. O. Box 679. NEW YORK.

Sophisticated and expensive baking equipment for the upscale home cook. *The Royal Baker and Pastry Cook*, compiled by G. Rudmani, 1877. Author's collection.

recipe in the book that does not use baking powder, the traditional "Wedding (or Bride) Cake." A dense, dark fruitcake made with, nuts, spices, and alcohol, this is a holdover from Renaissance Italy's *panforte*—literally, "strong bread." There are also recipes for what are now classic American foods: "Pumpkin Griddle Cakes," "Huckleberry (Blueberry) Griddle Cakes," "Sweet Potato Buns (Biscuits)," "Graham Crackers," "Ginger Snaps," "Lemon Meringue Pie," and "Strawberry Shortcake." A recipe for a cream cake with custard between the layers and a chocolate glaze on top gained fame when it became known as Boston cream pie.

All of the recipes might not have been new, but the cookbook itself was something completely new: a cookbook that was 100 percent advertising. In addition to the word "Royal" in almost every recipe, there are seven to eight lines of straight advertising text at the bottom of every page: "Do not use the goods of other manufacturers in these recipes"; "It is poor economy, in trying to save a few pennies on baking powder, to sacrifice your health"; "Yeast used for leavening purposes destroys the nutritive elements of flour"; and so forth. Royal also rang the dyspepsia bell repeatedly: "Indigestion, Sour Stomach, and Dyspepsia are often brought on by the use of Alum powder."[22]

Another novelty the cookbook advertised was that Royal was sold in "securely labeled tin cans." This was a scientific improvement over the old way of selling baking powder in bulk bins, or loose and wrapped in paper, where it "loses its strength."[23] The branding also aided advertising, because it differentiated Royal from all of the other baking powders on the market and from the generic baking powder in bins.

The Royal Baker and Pastry Cook spoke directly to the concerns of women who were skeptical of baking powder: "Many housekeepers are of the impression that baking powder is a chemical compound, dangerous to use; this is true of the cheap kinds which are mixed with the same ingredients used to adulterate Cream of Tartar."[24] This late nineteenth-century advertising says implicitly about Royal baking powder what became explicit in late twentieth-century advertising: "It costs more, but I'm worth it."[25]

The Royal Baker and Pastry Cook also provided testimonials. The inside front cover has three: from the state assayer of Massachusetts; from a professor of chemistry at the University of Pennsylvania; and from Henry A. Mott, a New York chemist with a PhD, and Royal's chemist. They all stated that they had analyzed Royal baking powder and found it "pure and wholesome." A fourth authority, Professor Charles F. Chandler, president of the newly created New York City Board of Health, is also quoted throughout the booklet, including at the bottom of the page. His most important statement appears at the front of the booklet, under the all-uppercase heading: "DO NOT USE CREAM OF TARTAR AND SODA." Royal did this to discredit its own ingredients in every other form. It says that Professor Chandler "found nearly all the Cream of Tartar sold by Grocers was adulterated from 80 to 90 per cent. with white clay (Terra Alba), Alum, and other hurtful substances."[26]

The cookbooklet also discredited all other leaveners, including yeast. At the Centennial Exposition in Philadelphia in 1876, the Fleischmann brothers had introduced a new style of Viennese yeast. This revolutionary yeast was not the liquid, sloppy slurry that women had made at home for centuries and had to take such pains to store, the one that easily went rancid. In the Fleischmann yeast, the liquid was removed and the yeast was compressed into small cakes. For the first time, bakers, both residential and commercial, had consistent, reliable, easily portable and storable baking yeast that was not made from bitter hops. This was a real threat to baking powder. The 1878 Royal cooking pamphlet used Horsford's 1861 *Theory and Art of Bread-Making* almost verbatim to discredit yeast and the evils of fermented bread. The first two recipes are for "Royal Unfermented Bread" and "Graham Unfermented Bread." The pamphlet also attacked Rumford indirectly, by disparaging baking powders made from "burntbones."

THE ROYAL BAKER AND PASTRY COOK. 9

66—Graham Flour Puffs.

1½ pints Graham flour, 1 teaspoonful salt, 2 large teaspoonfuls Royal Baking Powder, 2 eggs, and 1 pint milk.

Sift together Graham, salt, and powder; add the beaten eggs and milk; mix together into a smooth batter as for cup cake, half fill cold gem pans (fig. X), well greased, and bake in hot oven 10 minutes.

67—German Puffs.

1 pint flour, 2 tablespoonfuls sugar, a pinch of salt, 1½ teaspoonfuls Royal Baking Powder, 3 tablespoonfuls butter, 4 eggs, 2 oz. sweet almonds, 3 drops Royal extract bitter almonds, ½ pint cream, ½ cupful sultana raisins, ½ wineglass rum.

Rub the butter and sugar to a white, light cream; add the eggs (whole) 1 at a time, beating three or four minutes between each addition; blanch the almonds. (See Recipe No. 263).

Sift together flour, salt, and powder, which add to the butter, etc., with the almonds, raisins, extract of bitter almonds, cream, and rum. Mix the whole together into a smooth batter as for pound cake; two-thirds fill well greased cups; bake in a fairly hot oven 20 minutes; at the end of that time insert a straw gently, and if it comes out clean, they are ready; if any of the uncooked batter adheres to the straw, they must be set carefully back a few minutes longer.

68—Royal Oatmeal Puffs.

½ pint oatmeal, ½ pint Graham flour, ½ pint flour, 1 teaspoonful sugar, ½ teaspoonful salt, 2 teaspoonfuls Royal Baking Powder, 3 eggs, 1 pint milk.

Sift together oatmeal, Graham, flour, sugar, salt, and powder; add the beaten eggs and the milk; mix into a thin batter; half fill gem pans (fig. X), well greased and cold. Bake in good hot oven 10 or 12 minutes.

69—Flemish Waffles.

1½ pints flour, ½ teaspoonful salt, 2 tablespoonfuls sugar, 3 tablespoonfuls butter, 1½ teaspoonfuls Royal Baking Powder, 4 eggs, ½ pint thin cream, 1 teaspoonful each Royal extract cinnamon and vanilla.

Rub the butter and sugar to a white, light cream; add the eggs, one at a time, beating three or four minutes between each addition. Sift flour, salt, and powder together, which add to the butter, etc., with the vanilla, cinnamon, and thin cream. Mix into a smooth batter as for griddle cakes. Meanwhile, have the waffle-iron hot and carefully greased; pour enough batter in to fill the iron two-thirds full, shut it up and turn it over immediately; the iron should be hot without being too hot, and the waffles are to take 4 or 5 minutes. When ready, sift sugar over them, and serve on a napkin at once. (Bake in Fig. XIX.)

70—Soft Waffles.

1 quart flour, ½ teaspoonful salt, 1 teaspoonful sugar, 2 teaspoonfuls Royal Baking Powder, 1 large tablespoonful butter, 2 eggs, and 1½ pints milk.

Sift together flour, salt, sugar, and powder; rub in the butter cold; add the beaten eggs and milk; mix into a smooth consistent batter, that will run easily and limpid from the mouth of the pitcher. Have the waffle-iron hot, and carefully greased each time; fill it two-thirds full, and close it up; when brown turn over. Sift sugar on them and serve hot.

71—German Waffles.

1 quart flour, ½ teaspoonful salt, 3 tablespoonfuls sugar 2 large teaspoonfuls Royal Baking Powder, 2 tablespoonfuls lard, the rind of 1 lemon, grated, 1 teaspoonful Royal extract of cinnamon, 4 eggs, and 1 pint thin cream.

Sift together flour, sugar, salt, and powder; rub in the lard, cold; add the beaten eggs, lemon rind, extract, and milk. Mix into a smooth, rather thick batter. Bake in hot waffle-iron; serve with sugar flavored with lemon.

72—Scotch Short-Bread.

1½ pints flour, ½ teaspoonful salt, 4 tablespoonfuls sugar, 4 tablespoonfuls butter, 1 teaspoonful Royal Baking Powder, 3 eggs, 1 teacupful milk, 1 teaspoonful Royal extract of orange.

Sift together flour, sugar, salt, and powder; rub in the butter cold; add the beaten eggs, nearly all the milk, and the extract; mix into a smooth dough without much handling. Flour the board, turn out the dough, roll it with the rolling-pin to quarter inch in thickness; cut with a knife into shape of small envelopes lay them on a baking tin (fig. XIV), wash them over with the remainder of the milk, lay on each three large thin slices of citron and a few caraway seeds. Bake in moderate hot oven 20 minutes.

73—Royal Sally Lunns.

1 quart flour, 1 teaspoonful salt, 2 teaspoonfuls Royal Baking Powder, ½ cup of butter, 4 eggs, ½ pint milk.

Sift together flour, salt, and powder; rub in the butter cold; add the beaten eggs and milk; mix into a firm batter like cup cake, pour into two round cake tins, the size of pie plates; bake 25 minutes in a pretty hot oven, or until a straw thrust into them gently comes up free of dough.

74—Rusks.

1½ pints flour, ½ teaspoonful salt, 2 tablespoonfuls sugar, 2 teaspoonfuls Royal Baking Powder, 2 tablespoonfuls lard, 3 eggs, 1 teaspoonful each Royal extract nutmeg and cinnamon, ½ pint milk.

Sift together flour, salt, sugar, and powder; rub in the lard cold; add the milk, beaten eggs, and extracts. Mix into a dough soft enough to handle; flour the board, turn out the dough, give it a quick turn or two to complete its smoothness. Roll them under the hands into round balls the size of a small egg; lay them on a greased shallow cake pan (fig. XIII), put very close together; bake in moderately heated oven 30 minutes; when cold, sift sugar over them.

75—Scotch Scones.

1 quart flour, 1 teaspoonful sugar, ½ teaspoonful salt, 2 teaspoonfuls Royal Baking Powder, 1 large tablespoonful lard, 2 eggs, nearly 1 pint milk.

Sift together flour, sugar, salt, and powder; rub in the lard cold; add the beaten eggs and milk; mix into a dough smooth and just consistent enough to handle. Flour the board, turn out the dough, give it one or two quick kneadings to complete its quality; roll it out with rolling-pin to one-third inch in thickness, cut out with sharp knife into squares larger than soda crackers, fold each in half to form three-cornered pieces. Bake on a hot griddle about 8 or 10 minutes; brown on both sides.

☞ *It is poor economy, in trying to save a few pennies on baking powder, to sacrifice your health. Acid Phosphate of Lime (burntbones), Patent Cream Tartar, Alum, Terra Alba, and in fact, every cheap trashy substitute so nearly resemble a genuine baking powder that it is impossible for the housekeeper to distinguish the difference by the appearance. It is therefore of the utmost importance to get the original and well-known "Royal," the oldest and best, which has stood the test of years. Recommended by eminent physicians and chemists everywhere for its health-giving qualities, great strength, and absolute purity.*

The Royal Baker and Pastry Cook discredited all other baking powders (see paragraph at bottom). *The Royal Baker and Pastry Cook*, compiled by G. Rudmani, 1877. Author's collection.

The Royal Baker and Pastry Cook severed multiple cultural connections that earlier cookbooks had reinforced and repeatedly undermined female authority. First, it completely removed the female voice. Women in the new republic, Jacksonian, and antebellum America wrote cookbooks in their own recognizable voices, based on their experiences in their home kitchens. But the products of the Industrial Revolution—new chemical products, such as baking powder and flavoring extracts, and new factory-made kitchen equipment and tools—needed new authoritative voices. *The Royal Baker* removed the female voice and replaced it with the impersonal scientific and professional male authority of the mysterious and misspelled G. [Giuseppi] Rudmani, "Professor of New York Cooking School." The prestigious-sounding school had been founded the year before to teach immigrants living in tenements how to cook. The corporate voice does not advise, cajole, or commiserate. It imparts information that the consumer longs to hear and that makes the consumer want to buy the product. It relieves the consumer of responsibility.

Earlier, female-written cookbooks had provided not only recipes but also advice, information on household management, as well as the care and feeding of invalids, and moral instruction through food. All of the information was based on each author's own life experiences. Corporate cookbooks, on the other hand, contained no sections on inexpensive food or food for invalids. This separated the mother from her healing function and undermined her medical authority.[27]

Philosophical prescriptions and proscriptions were also gone. *The Royal Baker* broke the tradition of abstemiousness in Western culture that went back to Aristotle. In corporate cookbooks there is no mind-body split; there is only the body and its gratification, as quickly as possible. The olfactory senses of smell and taste, disdained by classical philosophers as belonging to the lower elements, are elevated and privileged in cookbooks.

The Royal Baker and Pastry Cook and other corporate cookbooks also severed religious connections to food. Food became secular. These cookbooks removed the temptation inherent in food since Eve's unilateral foray into pomology. *The Royal Baker and Pastry Cook* contained no caveats about overeating, no words of warning about the temptations of desserts or the virtues of self-control. Instead, it was possible to bake your cake and eat it, too—as often as you wanted and in as many different ways as you pleased. With culinary hedonism, could sexual hedonism be far behind? In a corporate cookbook, the only sin is using a product made by another company.

In addition to advertising in local newspapers and printing proprietary cookbooks, Royal placed an article in *Scientific American* magazine on November 16, 1878. The author of the article, Dr. Henry A. Mott, sang the Royal theme song: other baking powders were adulterated with alum; alum was dangerous; alum in bread was against the law in England; other cream of tartar baking powders were made with an inferior grade of cream of tartar; other baking powder manufacturers were only interested in "dollars and cents." Only Royal was made from the "pure grape."[28] Letters from Royal's competitors flooded the magazine. People in the trade knew that Mott had been Royal's chemist and wrote sardonically that they hoped he had been "liberally requited."[29]

However, Mott and Royal had a powerful ally in Ellen Swallow Richards, a professor of chemistry at the Massachusetts Institute of Technology (MIT). Richards had become interested in baking powder and wrote about it in 1882 in her book *The Chemistry of Cooking and Cleaning*. Richards came down firmly on the side of cream of tartar and against alum and soda as "possibly the most injurious" combination.[30] Richards passed this message on to the generations of women chemists she trained or influenced, either at MIT, through her writing, or as the first president of the Home Economics Association, founded in 1899 in Lake Placid, New York.

Trade Cards: "The Art Education of a Nation"

Unlike the black-and-white *Royal Baker and Pastry Cook*, trade cards were based on color images. Beginning in the 1860s, inexpensive color lithography made these 2 × 4 inch or 3 × 5 inch cards popular across the country.[31] They were visible signs of consumption even after the product had been consumed. "At the height of its popularity in the 1880s, the trade card was truly the most ubiquitous form of advertising in America," according to trade card historian Robert Jay.[32]

Lithographs depicted many of the same things that people post on the internet now: sentimental and humorous scenes of children and animals, flowers, landscapes, sunsets, and moonrises. Even though lithographers like Currier and Ives advertised that they were "Publishers of Cheap and Popular Pictures," their pictures still cost money.[33] Trade cards, however, were free. Children, especially girls, collected trade cards and put them into scrapbooks. Through play, this taught girls brand awareness and how to become consumers.[34]

Jay divides trade cards into five general themes: "patriotic imagery, the contrast between city and country, racial stereotypes, womanhood and the home, and . . . children."[35] Baking powder trade cards, too, displayed these themes. However, Jay fails to mention that often the women and children were sexualized. Naked cherubs cavorted across Rumford cards; women languished for baking powder.

Two foods mentioned frequently on baking powder trade cards are cakes and biscuits. Czar baking powder guaranteed it would make "Healthy and Delicious Cakes [and] Biscuits." Aunt Sally's baking powder made "splendid biscuits and cakes." Cleveland's made "the finest cake and biscuit." Redhead's mentions bread, biscuits, and cake. Words repeated on baking powder trade cards were "pure," "superior," "healthful," "nutritious," and "wholesome." The alum baking powders added "cheapest" and "uniform"—two important attributes of industrial food.

In addition to selling baking powder, trade cards cross-promoted products that were distributed by the company that made the baking powder or were sold in the same store. These were usually spices, extracts, and flavorings. Aunt Sally's promoted bluing, a laundry detergent.

The obverse of baking powder trade cards appealed to consumers in two ways: emotion and economics. The appeal to emotion was through "Reason Why" advertising. This was a new type of advertising created by C. W. Post of Post cereal. Previously, advertising simply described the product or announced that a product was available; "Reason Why" ads provided irrefutable reasons why the customer could not do without it and needed to buy it. However, the reasons were often emotional needs or vague statements. For example, the obverse of the Redhead's baking powder card said, "Reasons Why Redhead's Baking Powder Is Superior To All Others," and then listed them: it was uniform, never failed, "Keeps the household happy," and other vague statements.

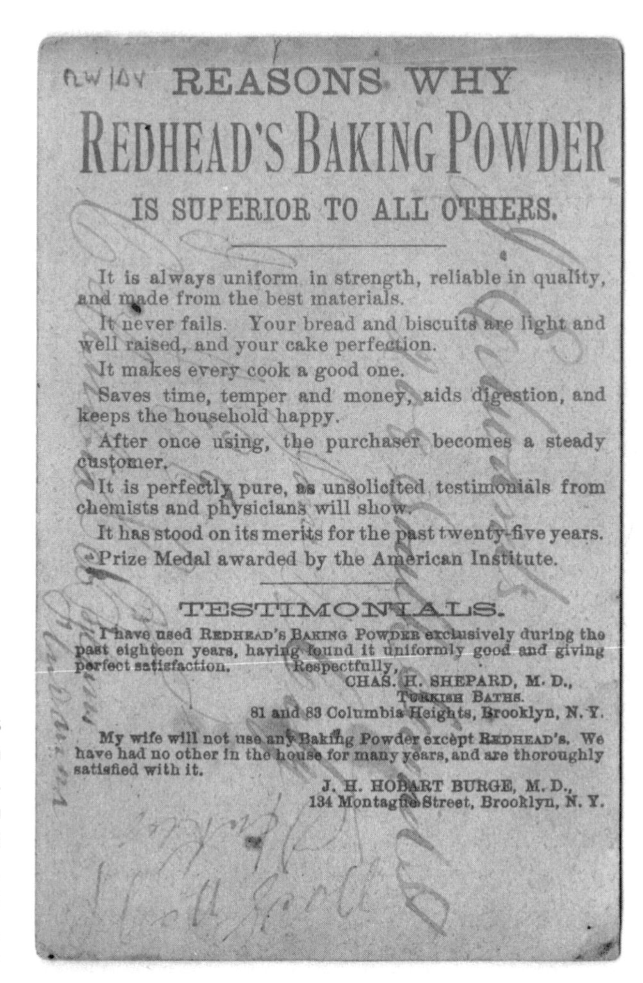

Redhead Baking Powder's "Reasons Why" was a new advertising concept. Nahum (Nach) Waxman Collection of Food and Culinary Trade Cards, Division of Rare and Manuscript Collections, Cornell University Library.

Testimonials were a type of "Reason Why" advertising, with the extra benefit that the person providing the reasons was an expert. For baking powder, the expert was usually a chemist or physician, the more eminent the better. Czar baking powder in New Haven, Connecticut, featured a testimonial from S. W. Johnson, professor of chemistry at Yale's Sheffield Scientific School. Sterling baking powder garnered a statement from the chemist for the U.S. Army. Dr. Price's card claimed that it "received the highest testimonials from the most eminent chemists in the United States," but did not name any. Patapsco did name its "eminent chemists and physicians" from Chicago, Boston, Philadelphia, Yale, and New York City's Bellevue Hospital. Redhead's baking powder card mentioned its "Prize Medal Awarded by the American Institute," with no information on what or where "the American Institute" was. Oddly, trade cards gave consumers many reasons why they should use baking powder but not what they should do with it. Baking powder trade cards rarely had recipes printed on them.

In addition to testimonials proving purity, economic incentives like gifts and refunds enticed customers. Babbitt baking powder gave away coupons with the purchase of each can. For four coupons from a one-pound can (thirty cents), eight coupons from a half-pound can (fifteen cents), or ten coupons from a quarter-pound can (ten cents), the purchaser would receive a free 14×28 inch panel picture from a selection of sixty designs. First-time purchasers of a pound of DeLand's baking powder could receive a reprint of an illustrated children's holiday book about two little girls named Wonder Eyes and What For. Some baking powders, like Union, promised simply a generic "present." An 1884 Cleveland's baking powder card promised a free cookbook to anyone who sent her address to Cleveland's offices at 81-83 Fulton Street in New York City.

Sometimes the consumer did not have to buy the baking powder to get a gift. Instead, she had to attend a demonstration of the product. Horsford's baking powder promised women in Kansas City, Missouri, a tasting of "biscuits, muffins, and gems" if they came to a baking exhibition. For one year, beginning on April 1, 1887, Rumford budgeted a hefty twelve thousand dollars for exhibitions and other advertising in Chicago.[36]

The promise of a refund provided a security net to the hesitant consumer. J. Monroe Taylor's baking powder offered a full refund if it did "not give entire satisfaction." Other refund offers were hyperbolic. In New York, Union baking powder offered a one-thousand-dollar reward "if proven impure, or adulterated, or short weight." Baltimore-based Sterling offered a thousand dollars to any chemist who could find "Alum, Bone Dust, or impurities of any kind" in its baking powder. Bone dust, of course, referred to phosphate baking powders like Rumford.

The new lithographic art also reflected the deeply ingrained racism in American society, which was found in other types of cards as well. In contrast to the Edenic images on its Caucasian "home sweet home" cards, Currier and Ives also printed the racist Darktown series. These cards depicted, supposedly humorously, the parallel universe inhabited by African Americans, Native Americans, and Asians. Inept black people dressed in rags failed at simple tasks and did stupid things that made them topsy-turvy. Often they were bested by white children, or by animals like pigs and dogs. What the Currier and Ives lithographs revealed was the soul-corroding ridicule to which African Americans were casually subjected on a daily basis.[37]

These stereotypical racist attitudes also permeated business trade cards. One baking powder trade card shows a young white girl in a kitchen supervising an African American man mixing dough in a bowl. He wears a white apron and white hat, but it is not a tall chef's hat. It is flat on top, like the hat an organ-grinder's monkey wears. The man's right leg is chained to the floor; he is her slave. The message of the card is clear: using this brand of baking powder is as good as having a slave. Another card shows a young black man carrying a giant box of Monroe's baking powder on his shoulders. He has an expression of terror on his face because he is staring at a white cat. This is supposed to be a humorous play on white people's superstitious fear of black cats. Aunt Sally's baking powder has the familiar black mammy figure holding a pan of biscuits. She says, "Dar's no use talking. Missus Vickery's Aunt Sally baking powder am de best for biscuits and cake."[38]

Trade cards declined at the end of the nineteenth century because advertisers found more efficient ways to reach a mass audience. (The notable exception was baseball cards.) The U.S. Post Office was responsible in large part for the democratization of advertising, along with the railroad. In the eighteenth century, mail was delivered to the town tavern; at the beginning of the nineteenth, to the general store. By the end of the nineteenth century, mail was delivered directly to private homes. With the advent of bulk mail, the post office shipped magazines for a flat rate. This meant there was no limit to the number of advertising pages. Magazines could and did contain hundreds of pages of advertising, which created profits for publishers. Catalogues, too, could be mailed directly to consumers. Sears and Montgomery Ward shipped catalogues and goods from Chicago throughout the country.

In spite of *The Royal Baker and Pastry Cook*, the trade cards, and the newspaper and other advertising, the baking powder companies that Royal attacked continued to do well and to proliferate. By 1884 Rumford was paying dividends of seven to eight dollars per share per month.[39] The company had expanded in multiple directions: physical plant, distribution, and product line. Factory and administrative functions were specialized and split into different buildings. The chemical factories were in multiple buildings along the river in East Providence. The research lab was relocated to a separate building in the city of Providence, along with the main offices and shipping and printing facilities.

In 1884 Rumford expanded the use of the phosphates in its baking powder to a new product: "the manufacture and sale of a beverage."[40] Soft drinks were a popular

This racist trade card equates baking powder with having a slave in the kitchen. Nahum (Nach) Waxman Collection of Food and Culinary Trade Cards, Division of Rare and Manuscript Collections, Cornell University Library.

new food, helped by inexpensive sugar and a growing temperance movement. There were no controlled substances, so if a liquid in a bottle had a high alcohol content, was it a medicine or an alcoholic beverage? It depended on when it was manufactured and who made the decision. In 1883 the commissioner of the Internal Revenue Service ruled that a beverage that contained alcohol was medicine. In 1905, with the public increasingly vociferous about the evils of alcohol, the new commissioner decided that beverages that contained alcohol were liquor.[41]

Soft drinks, too, could claim anything, and did. Even if beverages did not contain alcohol, they could contain drugs and make medicinal claims, because the line between "legitimate and illegitimate forms of consumption" was not clearly defined.[42] Hires Root Beer had been purifying blood since 1876. Beginning in 1885 in Texas, Dr Pepper aided digestion; in Lowell, Massachusetts, Moxie Nerve Food promised to cure nervousness and paralysis. In 1885 Atlanta physician-pharmacist John Stith Pemberton sought to break his morphine addiction by concocting a beverage steeped in stimulants. He called it Pemberton's French Wine

Coca. The following year, he changed the formula and renamed it after two of the stimulants it contained: Coca-Cola.[43]

Horsford's advertising sought to distance his beverage from the other soft drinks. Its temperance status was in its name: Horsford's Acid Phosphate (Non-Alcoholic). It also contained no drugs. Horsford attempted to legitimize his product by connecting it to his stature as a scientist. Horsford's Acid Phosphate was "not a compounded patent medicine, but a scientific preparation recommended and prescribed by physicians of all schools." Nevertheless, its claims were the same as those of other soft drinks. Just a teaspoonful added to a "tumbler of water" and sweetened to taste made "a delicious drink" that was good for a wide range of ailments: indigestion and dyspepsia, nervousness, headache, "tired brain," and weakened energy. It also claimed to be effective against both exhaustion and sleeplessness.[44]

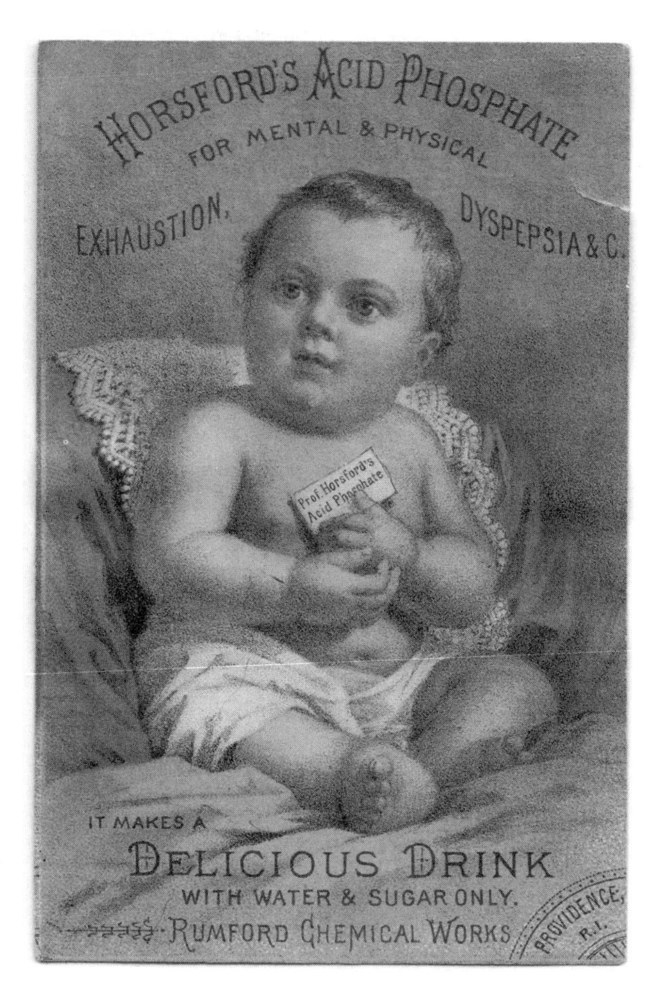

Babies were popular nineteenth-century advertising icons. Author's collection.

In the 1880s Rumford extended its prosperity to its employees. The company built roads and housing, planted trees, and established committees to look into creating a library for employees and a cemetery. The firm paid hospital bills for an employee with almost thirty years of service.[45] On March 19, 1886, the board of directors voted unanimously to give annual bonuses that rewarded longevity:

> To those whose wages amount to $1200 per year or less the annuity shall be:
>
> 10 per cent of the wages earned where the continuous service has been 10 years, and less than 15 years.
>
> 15 per cent where the service has been 15 yrs and less than 20
>
> 20 per cent where the service has been 20 yrs and less than 25
>
> 25 per cent where the service has been 25 yrs and less than 30
>
> To those whose wages amount to more than $1200. per yr, the annuity shall be one half of the above per centages for corresponding term of service.

Rumford's largesse also extended to women who left their jobs to get married but who had worked less than ten years. In keeping with common practice at the time, only single women worked. When a woman got married, her place was in the home caring for her husband and children. The money for the wedding presents to these women did not come from Rumford; it came personally from Eben Horsford, the father of five daughters.[46]

Horsford also strongly supported female education and was closely involved with Wellesley College, founded in 1870, which his daughters attended. He endowed the college's library, donated funds for scientific equipment and for sabbaticals, which he stipulated were for female faculty only and which had to be taken abroad.[47] For his contributions to the college, Horsford was made an honorary member of the Wellesley class of 1886.[48]

The Baking Powder War and Native Americans

Baking powder companies also fought for lucrative government contracts with the military and on Native American reservations. Dr. Henry Mott, a chemist who had analyzed baking powders for the United States Indian Commission, sued Jabez Burns, publisher of *The Spice Mill*, a grocery trade publication. At issue were remarks that Burns had printed about why Mott chose more expensive cream of tartar baking powder over alum baking powders for use on Native American reservations. The court found Burns guilty and fined him eight thousand dollars. Mott said that Americans should avoid buying baking powder sold loose or in bulk, because it was likely to be adulterated, and that "the label and trade-mark of a well-known and responsible manufacturer . . . is the best protection the public can have." This was front page, above-the-fold news in the *Brooklyn Daily Eagle*, under the uppercase heading "CONDEMNED. Alum Baking Powders in Court" and was reprinted in newspapers and periodicals across the country.[49] Mott later provided commercial endorsements for Dr. Price's cream of tartar baking powder.[50]

The number of Native Americans on reservations increased after the annihilation of Custer's Seventh Cavalry at the Little Big Horn in Montana in June 1876 made subduing Native Americans a priority. Food was part of the forced assimilation to which the American government and Christian missionaries subjected Native Americans, along with individual ownership of land, Christianity, and Western clothing. To Caucasians, baking powder use on Native American reservations was a sign of assimilation and civilization. By the 1880s, Native Americans on reservations in Idaho received a weekly ration of baking powder, along with flour and beef.[51] In Dakota Territory, Christian missionaries, appalled that the Arickaree, Mandan, and Gros Ventres tribes persisted in native practices like polygamy and scaffold burials (burying the dead in trees), regarded the use of baking powder as a positive sign: "They are eagerly obtaining from the Government such comforts of civilization as they can—reapers, cooking-stoves, baking-powder."[52]

What the Native Americans made with the baking powder was fry bread, a simple fritter made with nonperishable ingredients and deep fried. Fry bread became a dietary staple, as weighted with symbolism for Native Americans as bread is in European and American culture. But what it symbolized had nothing to do with religion. According to food historian Alice Ross, fry bread is "the most important of the foods of the pan-Indian movement and the symbol of intertribal unity." This is ironic because none of the ingredients in fry bread except salt are indigenous to any Native American culture. Neither is the technique. The intertribal unity is one of subjection.

NATIVE AMERICAN FRY BREAD

3 cups flour, either all white or half whole wheat
1⅓ cups warm water
1¼ teaspoons baking powder
½ teaspoon salt
Lard or oil

Mix flour, baking powder and salt. Add warm water and knead until dough is soft but not sticky. Stretch and pat dough until thin. Tear off one piece at a time, poke a hole through the middle and drop into kettle of sizzling hot lard or cooking oil. Brown on both sides. Serve hot.[53]

The color, consistency, and taste of wheat fry bread were very different from what Native Americans made with corn. Wheat is white or light brown. Americans standardized corn as yellow or white, with cobs of relatively uniform size so that the kernels could be removed by machine and canned. But Native American corn comes in vibrant rainbow colors: purple, scarlet, orange, yellow, mottled, blue, brown. It is with deep purple-blue corn that the Hopi make piki bread, as thin as tissue paper. Unlike fry bread, piki bread takes a great deal of skill to make. The baking stone on which it is made, like the knowledge of how to make piki bread, is passed from mother to daughter. The griddle is seasoned with oil made from sunflower or squash seeds. Then the batter is spread on the hot stone with the hand and cooks almost instantly. The translucent sheets are rolled up and used to soak up sauces or stews.[54]

Food historians agree that fry bread was a nineteenth-century innovation in Native American cooking but disagree about its origins. The Spanish and French had introduced Native Americans to wheat bread and to its use in the Christian sacrament. Some food historians see in fry bread the French influence of sweet yeast-risen bread. Others see a Spanish hand in the deep-frying technique that is also used in wheat sweets like churros and sopapillas.[55] However, the ingredients, especially baking powder, tell the truth: fry bread is a product of Native American incarceration and subjugation. Modern Native American activists like Susan Harjo want a ban on fry bread because it is not indigenous and because they believe it contributes to diabetes, which is prevalent among Native Americans.[56]

The shift from corn to wheat in Native American cooking was not benign. It was not a simple substitution of ingredients but had far-reaching religious and cultural implications. Corn was sacred in Native American cultures: deities, rituals, and festivals were connected to it. Separation of Native American women from corn disrupted their connections to religion and their identification with powerful female deities. It also changed women's connections to one another, because grinding corn was an hours-long communal activity. Fry bread is fast food.

Baking Powder and Scandinavian Immigrants

Baking powder use was an indication of assimilation not only among Native Americans but in immigrant populations as well. The dual-language *Swedish-American Cookbook*, first published in 1882, was an agent for assimilation through food. It also reveals how much Scandinavians in America had already assimilated and how deeply and quickly cream of tartar baking powder propaganda had spread even into immigrant cultures. The book is set in two side-by-side columns, with a recipe in Swedish on the left and the translation in English on the right. Many English words have already migrated to the Swedish side of the book, especially the words for measurements: "ounce," "cup," "gill," "pint," "quart," "gallon," and "handful." Words for many ingredients, too, are in English: "Hamburger-Steak," "Roast Biff [*sic*]," "Indian meal," "graham," "molasses," "lard," "soda," "cream of tartar," and "baking powder." Most of the breads are leavened with yeast, but there are also American-style recipes that use baking powder, among them baking powder biscuits, buckwheat cakes, and waffles. Even Vienna rolls have become Americanized and leavened with baking powder.

It is clear from the proportions of flour to baking powder that these recipes are for cream of tartar baking powder. For the anonymous author of the *Swedish-American Cookbook*, baking powder comes with a caveat: "Soda, saleratus, cream of tartar, and baking powder, as found in the American market, are often adulterated through mixture with terra alba or white sand. To test them, put a teaspoonful in a glass of water; if pure, it will dissolve, otherwise there will be a gathering at the bottom of the glass. Some baking powders contain alum and should not be used, being very hazardous to your health."[57]

Nevertheless, in the 1880s sodium aluminum sulfate gained ground in American markets because it was cheaper and stronger than cream of tartar. Sodium aluminum sulfate cost three cents per pound; cream of tartar cost thirty. Also, less sodium aluminum sulfate was necessary to leaven the same amount of flour.[58] This was a serious threat to Royal. They increased their advertising budget in the early 1880s when Henry La Fetra joined the company to manage the advertising department under Ziegler's supervision. By 1888 Royal had between eight thousand and nine thousand contracts with five to six thousand newspapers in the United States, England, Scotland, Canada, South America, Mexico, Africa, Australia, and the West Indies.[59]

As baking powder use increased, women had to learn to bake in new ways as they made the transition from yeast-risen breads to baking powder breads and cakes. Baking powder companies taught women through corporate cookbooks, but women continued to educate themselves and each other about the new leaveners through cookbooks that they wrote. Exactly how this transition occurred is evident in the *White House Cook Book*, published in 1887. The author, Mrs. F. S. Gillette, presents two recipes for the same breadstuff but with different leaveners: fermented with yeast, and unfermented with baking powder. The baking powder recipes are not in a separate section of the book; they immediately follow the yeast recipe for the same bread, so readers can compare. Gillette presents recipes for standards like graham bread, Boston brown bread, Parker House rolls, Sally Lunn, rusks, waffles, and griddle cakes, among others, in this way. For some of the recipes, like the one for Boston brown bread, Gillette points out the difference between the uses of baking soda and baking powder. If the recipe includes an acidic ingredient like sour milk, baking soda will react with it and cause it to rise. With nonacidic sweet milk, baking powder is called for.[60]

Recipes in the *White House Cook Book* show that by 1887 there had been a huge leap in the use of non-yeast leaveners. They outnumbered yeast by three to one. The cream of tartar and baking soda combination remained stagnant at 13 percent, the same percentage as in the *National Cookery Book* in 1876. Saleratus, which had been in almost 10 percent of recipes in 1876, was now completely gone. The

232 *BREAD.*

GRAHAM BREAD. (Unfermented.)

STIR together three heaping teaspoonfuls of baking powder, three cups of Graham flour and one cup of white flour; then add a large teaspoonful of salt and half a cup of sugar. Mix all thoroughly with milk or water into as stiff a batter as can be stirred with a spoon. If water is used, a lump of butter as large as a walnut may be melted and stirred into it. Bake immediately in well-greased pans.

BOSTON BROWN BREAD.

ONE pint of rye flour, one quart of corn meal, one teacupful of Graham flour, all fresh; half a teacupful of molasses or brown sugar, a teaspoonful of salt, and two-thirds of a teacupful of home-made yeast. Mix into as stiff a dough as can be stirred with a spoon, using warm water for wetting. Let it rise several hours, or over night; in the morning, or when light, add a teaspoonful of soda dissolved in a spoonful of warm water; beat it well and turn it into well-greased, deep bread-pans, and let it rise again. Bake in a *moderate* oven from three to four hours.

Palmer House, Chicago.

BOSTON BROWN BREAD. (Unfermented.)

ONE cupful of rye flour, two cupfuls of corn meal, one cupful of white flour, half a teacupful of molasses or sugar, a teaspoonful of salt. Stir all together *thoroughly*, and wet up with sour milk; then add a level teaspoonful of soda dissolved in a tablespoonful of water. The same can be made of sweet milk by substituting baking powder for soda. The batter to be stirred as thick as can be with a spoon, and turned into well-greased pans.

Mrs. Gillette's *White House Cookbook* helped to educate women in the transition from yeast (fermented) to baking powder (unfermented) baked goods. Library of Congress.

astounding increase was in commercial baking powder, which accounted for 40 percent of all leaveners and was used in 43 percent of cakes. Yeast had almost disappeared from cakes, except in the vestigial "Election Cake" (see appendix table A-5).[61]

In the little more than thirty years since Horsford had received his first patent in 1856, baking powder had grown into a multimillion-dollar business with hundreds of companies. Royal had grown phenomenally from its start in a small Midwestern pharmacy in 1866 to a New York City company selling throughout the United States and in multiple foreign markets. However, by the end of the 1880s, the cream of tartar companies, outnumbered by new alum baking powder companies, found their share of the total market shrinking. In the next phase of the baking powder wars, the cream of tartar companies would turn on each other.

CHAPTER 5

The Cream of Tartar Wars

Battle Royal, 1888–1899

Don't be deceived by these . . . articles in the newspapers. They are advertisements, prepared and paid for by the Royal Baking Powder Co.

—Cleveland Baking Powder Ad, 1890s

[While] . . . the cream of tartar interests . . . were busily exposing each other's defects . . . the entire south was irrevocably lost to cream of tartar and gained by the alum interests.

—American Baking Powder Association, 1904

A century after Amelia Simmons threw down the chemical leavening gauntlet in *American Cookery* in 1796, Americans were buying almost 120 million pounds of baking powder a year.[1] Bread was still "the most important article of food" in the American diet, but baking powder had changed how breadstuffs were made and created new kinds of breads, such as baking powder biscuits.[2] Baking powder had also become the default leavener in desserts, and Americans were making more desserts: "[Forty-two] percent of the recipes provided by the *Ladies' Home Journal* from 1884 through 1912 were for desserts."[3] Hundreds of baking powder companies vied for a share of this huge market, which continued to spread into new immigrant and ethnic communities in the United States, as well as into international markets. Competition was especially fierce among the cream of tartar companies, because that 120-million-pound figure was lopsided: almost 100 million of those pounds were alum baking powder; only a shrinking 20 million were cream of tartar.[4]

As competition for the national market increased, there were wars between companies that were local and national, rural and urban, small and large. The battle for profits was exacerbated by the recession that began in 1893 and new competitors who entered the market, as barriers to entry remained low. America's transportation and communications infrastructure removed barriers at the local level by allowing small businesses to purchase products from remote areas inexpensively. A family grocery in any small town could enter the baking powder business with inexpensive chemicals mined in the United States, along with tin cans, labels, and shelf space.

During the last twelve years of the nineteenth century, there were three baking powder wars. Royal was at the center of all three. Each war illustrates a different type of business war in the United States. That they all occurred at roughly the same time indicates how rapidly business moved in Gilded Age America. The first baking powder war was within Royal, for control of the company. The second was among the three leading cream of tartar companies. The third was the ongoing war involving the cream of tartar companies and the phosphate companies like Rumford, both of which were resisting the newer alum companies that were springing up everywhere, especially in the Midwest.

On a fundamental level, the wars were between old and new ways of conducting business, between European and American philosophies. Business historian Richard Tedlow defined the quintessential American style of business: "The strategy of profit through volume—selling many units at low margins rather than few units at high margins—historically has been the distinctive signature of the American approach to marketing. By making products available to the masses all over the nation—by democratizing consumption—the mass marketer did something profoundly American."[5] The alum baking powders, low-priced and widely available, followed this American model. However, Royal, dependent on the imported European wine by-product argol for its main ingredient, had difficulty with this model. Royal was selling on the European model of restricted consumption based on class: selling at low volume, high price, and high margin. It was inevitable that these two different modes of conducting business would come into conflict.

The first cream of tartar war in the last decade of the nineteenth century was for control of Royal Baking Powder. Royal had begun in 1866 as a two-man partnership, the most common type of business in antebellum America. It became a corporation in 1868. Royal's success was phenomenal. It was grossing close to $3 million annually by 1888. In the preceding eleven years, its total sales had increased almost eightfold, from approximately $350,000 to almost $2.7 million. The net profit margin had increased even more, thirteenfold. Its principals plunged into in a battle for control of the profits and the company.[6]

The Hoagland family, the original founders of the company, felt that they did all the work while William Ziegler got too much money. Ziegler felt the same about the Hoaglands. He also accused the Hoaglands of being corrupt, of taking kickbacks, and of using company funds to bribe the New York State legislature to influence legislation.[7]

The Hoaglands wanted Ziegler gone and repeatedly tried to buy him out. But Ziegler had been a trustee of the company from its inception and had increased his initial 140 shares of stock to 425, more than one-quarter of the 1,600 shares outstanding.[8] When Ziegler refused to sell his stock, the Hoaglands determined to force him out, and they outnumbered him. In the January 1887 annual stockholders' meeting, they did not reelect Ziegler as a trustee but did elect the Hoaglands—brothers Joseph and Cornelius, and Joseph's son Raymond.[9] Then they reduced Ziegler's profits by raising their own salaries. Until January 1888 the officers' salaries had been a nominal eighteen hundred dollars per year. Then the trustees voted in new salaries: Joseph C. Hoagland, president, fifty thousand dollars; C. N. Hoagland, vice president, thirty thousand dollars. Raymond Hoagland, twenty-one, was treasurer and made six thousand dollars. William M. Hoagland, C. N.'s nephew, was secretary and made three thousand dollars.[10] On January 23, 1888, Ziegler sued the Hoaglands and Royal.[11]

Ziegler was not left penniless, however. The *New York Times* reported during the trial that "Mr. Ziegler's otherwise good memory deserted him and he declared he did not know whether or not he had received $1,250,000 in dividends."[12] The judge sided with Ziegler and ruled that the salaries be cut. Joseph Hoagland's fifty-thousand-dollar salary was reduced to fifteen thousand; C. N.'s, from thirty thousand to ten thousand dollars; and Raymond's from six thousand to four thousand.[13]

The battle for control of Royal was just the beginning of a war for control of the cream of tartar share of the market. In the 1890s the three major cream of tartar

Table 5-1. Royal Baking Powder Company, Sales Profits, and Advertising, 1876–1887

	Total Sales	Net Profits	Advertising Expenditures	Advertising as Percentage of Profits
1876	$349,567	$55,810	$17,648	31.62
1877	346,945	61,708	22,529	36.50
1878	447,513	*68,770	*68,638	99.80
1879	646,133	143,306	86,713	60.50
1880	884,343	202,832	74,669	36.81
1881	1,175,350	308,820	97,666	31.62
1882	1,486,492	390,739	116,947	29.92
1883	1,895,434	511,303	149,224	29.18
1884	2,119,062	534,980	220,082	41.13
1885	2,213,288	564,272	289,566	51.31
1886	2,434,575	682,254	251,623	36.88
1887	2,657,987	725,162	296,085	40.83

Source: *Ziegler v. Hoagland,* BPC, 1, sales/profits, 612; advertising, 610–11.
* These are the amounts listed.

baking powder companies waged a war among themselves and changed hands in a stunning series of revolving-door deals. The precipitating factor was that Ziegler finally sold his interest in Royal to the Hoaglands for four million dollars. He turned around and used the money to buy Royal's competitor, the Chicago-based Dr. Price.[14] By 1895 Joseph Hoagland was sole owner of Royal; another Hoagland controlled Cleveland Baking Powder, the third major cream of tartar baking powder. The war was not just business; it was personal between Ziegler and the Hoaglands, and then between the Hoaglands.

In the war among the three major cream of tartar companies, Royal, Dr. Price, and Cleveland faced two major challenges. First, they had the ongoing struggle to convince housewives to buy commercial baking powder and to spend more for it than the baking powders they had been concocting in their own kitchens for years. Second, each cream of tartar company had to differentiate its product from the other cream of tartar baking powders, although they were essentially the same.

To prove that commercial cream of tartar baking powders were better than homemade, Royal set out to convince housewives that the baking powders they had been mixing in their kitchens were inferior. Royal continued the advertising, testimonials, and instructions in cookbooks to inform consumers that baking powder created by professionals gave "much better results than cream of tartar and soda hastily put together by an amateur." Also, it claimed that the cream of tartar used in commercial baking powders was of a higher quality than what women could purchase from their local druggist, and it was free from adulterants.[15]

The cream of tartar baking powders could not use price to differentiate their products, because they all cost more than alum baking powders. Royal's advertising acknowledged that it cost more but also said that the intelligent woman was willing to pay the price for a better product that showed her discernment. Cleveland also advertised that its cream of tartar baking powder "costs a few cents a pound more than alum powders, but it is worth more."[16]

Because they were more expensive than alum baking powders, all three cream of tartar baking powders had to aim at the same market: aspirational consumers. Royal's name and advertising encouraged this. The words "Royal Baking Powder" were gold on a red label, two colors historically associated with royalty. The Dr. Price name had the cachet of medical authority and security that the educated woman would value. All the cream of tartar powders claimed they were "pure." These were not great differences.

Cleveland attacked head-on and revealed Royal's advertising-as-straight-news practices. In a brochure subtitled "A 'Royal' Fraud Exposed," Cleveland reprinted some of Royal's planted articles. The Royal articles cited authorities, including a British professor, an unnamed French chemist, and an encyclopedia, that lauded the superiority of ammonia, which Royal contained. Cleveland warned readers,

"Don't be deceived by these and similar articles in the newspapers. They are advertisements, prepared and paid for by the Royal Baking Powder Co."[17]

While the cream of tartar companies fought each other, their phosphate and alum competition took advantage of the opportunity to expand in number and in market share. In 1892 Rumford bought two brick warehouses in Richmond, Virginia, to expand and facilitate distribution in the South.[18] However, the Rumford phosphate formula was also in the minority. The majority of the new baking powders were made from a mineral, sodium aluminum sulfate.

On March 2, 1899, William Ziegler ended the internecine cream of tartar war. Already the owner of Dr. Price's, Ziegler also brought pressure on Royal by buying the Tartar Chemical Company in Jersey City, which supplied Royal with cream of tartar.[19] Ziegler combined the three major cream of tartar baking powder companies—Royal, Dr. Price, and Cleveland—and created the Royal Baking Powder Corporation in New Jersey, which had changed its laws to make them more favorable to corporations. Royal was just one of 63 mergers that took place in 1899, the peak year for consolidations. Between 1895 and 1904, an unprecedented 157 trusts were formed in what became known as the Great Merger Movement. Like Royal, the majority of these mergers were horizontal, the combination of several companies in the same business. Unlike Royal, the majority of the mergers resulted in control of 40 percent of the market or more.[20]

The Royal trust was capitalized at $20 million, half of it preferred stock at 6 percent and half common stock.[21] The preferred stock was offered for sale to the public; the common stock was privately held. At the time, $20 million was an enormous sum. For example, railroads built in the 1850s, with massive investments in rolling stock, were capitalized at $5 million.[22] The Ford Motor Company's capitalization at its inception in 1903 was $150,000.[23] The bank for Royal's transaction was the "United States Mortgage & Trust Company . . . organized to control the baking powder business of the United States."[24]

The alum baking powder companies questioned the value of the stock; they believed it was watered. They also believed that Ziegler held all of the common stock, valued at $10 million. Royal admitted that perhaps 9 percent had been "given to friends."[25] This still would have left Ziegler with 91 percent of the common stock, valued at more than $9 million.

The Royal trust, like other trusts formed by the mergers that took place at the end of the nineteenth century, faced two major challenges: (1) "its costs of production relative to those of its competitors" and (2) its ability "to erect barriers to new competition."[26] Royal was losing on both counts. Cream of tartar baking powders, dependent on European imports of argol, cost more to produce than alum baking powders, which used American minerals. As for barriers to production, alum baking powder companies were proliferating. Unable to lower its costs, Royal needed to find a way not only to prevent new companies from entering the market but also to reduce or eliminate the hundreds of alum baking powder companies already in existence.

With Royal capitalized at $20 million, Ziegler had an annual advertising budget of $700,000. He finally had the financial means to do what Royal had been advocating since *The Royal Baker and Pastry Cook* in 1877: erect the ultimate barrier against alum baking powders and have them declared illegal because they were toxic. Ziegler had many options. He had only to decide on the best plan and proceed with it.

American passions over toxic foods and the people who produced them had been inflamed during the Spanish-American War in 1898, when "embalmed beef" caused a scandal. General Nelson Miles said, "I believe that 3,000 United States soldiers lost their lives because of adulterated, impure poisonous meat."[27] Ziegler's strategy was to have state legislatures pass pure food laws that would outlaw alum—not the baking powders, just the one ingredient—on the grounds that it was toxic.

Royal's advertising blamed alum for many ills, including "Why Women Are Nervous." It claimed that "the frequent cases of nervous prostration or utter collapse of the nervous system under which women 'go all to pieces'" were caused by alum. Although it caused "no visible symptoms" immediately, "loss of appetite and other alimentary disturbances, and finally a serious prostration of the whole nervous system" soon followed. As with other advertisements in the baking powder war, no specific authorities or studies were cited.[28]

It would take only one key state to pass an anti-alum law that would start a grassroots movement propelled by consumer awareness and pressure. The law could be used as an example to pressure other states. It was the same tactic Royal used with advertising: one newspaper story about deaths or illness attributed to baking powder, even if it was not true, would be carried by other newspapers. Royal's contracts with newspapers across the United States would ensure dissemination of the news that states were passing anti-alum laws. With a foothold in the states, Royal would be in an excellent position to win the coming war over pure food at the national level. If the states passed anti-alum laws, then the federal pure food law would have to exclude alum, too.

Royal's strategy to have its competitors' products declared illegal was not original. Nineteenth-century industrialization spawned wars between older, traditional, expensive foods and new, industrial, cheaper foods. In addition to baking powder, prime movers in the chocolate, whiskey, butter, and sugar industries who found themselves challenged by innovators used science to discredit their competitors' products and used law and public policy to try to drive their competition out of business. They advertised and lobbied and were tenacious in finding new ways to wage trade wars. All of these innovations changed not only their own industries, as alum had changed baking powder, but American eating habits as well.

The chocolate wars, like the baking powder wars, centered on the use of a new chemical—and it was the same chemical. In the 1860s, Dutch chemist Coenraad van Houten added potassium carbonate (pearlash) to chocolate. This heightened the chocolate flavor, darkened the color, and made chocolate more easily soluble in liquids like milk and water. Throughout Europe, consumers clamored for the convenient new "Dutch process" cocoa. This cut into the business of Cadbury, British chocolate manufacturers. Cadbury's advertising, like Royal's, equated purity with quality, so Cadbury could not compete by producing a similar cocoa. It could, however, fight on the grounds that van Houten added "alkalis and other injurious colouring matter," which made its cocoa "dangerous and objectionable." As in the baking powder war, the majority of consumers ignored the ensuing battle of the scientific experts. By the 1890s, 50 percent of British chocolate purchases went to van Houten.[29]

The war between straight and blended whiskey was similar to the chocolate war: both involved chemical additions to existing products.[30] Straight whiskey was aged in wood, then bottled. Blended whiskey, also called "rectified," used straight whiskey as a base; diluted it with cheaper, colorless grain alcohol; and added coloring and flavorings. Like van Houten's chocolate, less expensive blended whiskey was a runaway success, outselling straight whiskey more than three to one. Straight whiskey wanted to put blended whiskey out of business; blended whiskey fought back, and its business increased.[31]

The butter and oleomargarine war, too, was long and bitter. However, it was different from the baking powder, chocolate, and whiskey wars because although it involved products that served the same function, they had no ingredients in common. Oleomargarine was an inexpensive butter substitute patented in 1869 in France. Made from beef fat, it was a "pure dead white," so producers dyed it yellow to look like butter. In 1884 the dairy industry in New York state succeeded in getting a law passed that made manufacture or sale of butter substitutes illegal. The New York Supreme Court overturned the law. However, a law controlling the color of oleomargarine to differentiate it from butter was successful. "Bogus butter" could be any color as long as it was not yellow. What the dairy industry really wanted, but did not get, was oleomargarine dyed an unappetizing black, purple, pink, blue, or red.

Attempts to remove oleomargarine's chief attraction, price, by taxing it heavily also failed. Congressmen did not want to start down the slippery slope of setting public policy by taxing foods out of existence on oleomargarine-greased skids.[32] Ironically, in 1911 Procter & Gamble borrowed a page from its Ivory soap advertising and turned whiteness into a marketing asset—purity—when it created its cottonseed oil shortening, Crisco.[33]

The sugar–saccharin war, too, was between a food and a substitute. Saccharin was cheaper than sugar and, because it was synthetic, was not subject to the vagaries of weather, shipping, labor, and politics, as sugar was. The issue here, however, was one of fraud: people thought they were getting sugar but were not. The emotional controversy over saccharin illustrates how lost the baking powder war and the controversy over alum have become. Recent scholarship on the sugar–saccharin dispute says, "It is difficult to imagine, however, a similarly evocative campaign against alum."[34]

Ziegler learned from all of these wars. The near miss in the butter industry proved that a law prohibiting the sale of a specific food would not work. But what if something in the food were poisonous? That one ingredient could be declared illegal, which would force the product to be removed from the market.

What was at stake in the baking powder war? In 1900 there were approximately 534 companies manufacturing baking powder in the United States.[35] The American Baking Powder Association estimated that to shift from alum to higher-priced cream of tartar baking powder—that is, if Royal had a monopoly on all baking powder—would cost the American people "*something over $60,000,000 per annum. This stupendous volume of business would go directly to the Royal Baking Powder Company*."[36]

Royal also continued its war against other baking powders. It used puffery to discredit a common practice among salespeople at the time: door-to-door demonstrations. Royal attacked its competitors' salespeople as "a band of baking powder tramps." It characterized the "baking powder itinerants" as dangerous to the public health and claimed that alum baking powders had to stoop to this method of selling because they "have failed to find purchasers through legitimate means."[37]

ARREST HIM ON SIGHT

There is no greater or more dangerous traveling nuisance in the country than the fellow who goes from house to house in town or country leaving sample packages of patent medicines or foods in houses, or porches, doorsteps, or in yards. This thing has been done frequently in this town during the past year. Only a few days ago a lady used a package of baking powder left at her home with the result that all who ate of the food were made sick. Numerous cases of illness from a similar cause have been reported from various parts of the country in the past few months. Our constable or policeman should have strict orders to arrest on sight any person distributing medicine or food samples in this way. It is too dangerous a practice to be tolerated.[38]

At the same time that Royal was discrediting other baking powder companies' sales teams, it was sending out its own. Several women would go out under the aegis of a salesman. The saleswomen told housewives, "We represent no baking powder firm nor do we use or recommend the use of any particular brand of powder."[39] Instead, they said, they had been sent by William E. Mason of Illinois, an anti-alum U.S. senator who presided over the committee that was holding hearings on pure food. As government representatives, they said, their mission in visiting the home was "to inspect the baking powder used and show which powders contain alum and which are made of cream of tartar. We simply desire to demonstrate which powders are pure and which are not."[40]

During these tests, Royal cast the ammonia in its own baking powder in a positive light *because* it smelled bad: "a baking powder that would give off no gas when subjected to heat would be without leavening power."[41] This is true, but the gas did not have to smell like urine. The sales team then did demonstrations that "proved" that every baking powder except Royal was impure.

In an article titled "Burn the Samples," Royal claimed that alum baking powders cost "less than four cents a pound, so that they yield an enormous profit, which enables their makers to give them away in large quantities, or accompany their sale with all kinds of presents." In other words, all baking powder companies except Royal were motivated by greed. In the same article, Royal talked about an adulterated baking powder from Providence, a clear attack on Rumford.[42] Royal repeatedly refers to itself as "a high class cream of tartar powder," while alum baking powders are "cheap" and sold by unscrupulous people. Royal's ads also appealed to housewives' vanity, because "any intelligent person" would be able to distinguish the inferior alum powders.[43]

Baking powder companies appealed increasingly to the public through newspaper advertising but still advertised directly to the grocery trade. In 1898 the Dr. Price company mailed a pamphlet to grocers across Wisconsin to alert them to a state law that went into effect on January 1, 1898. The law required all alum baking powders to state clearly on the label: "This baking powder contains alum." The article then listed the names of 387 alum baking powders in alphabetical order, from Acme to Zipp's.[44] The pamphlet was mailed to grocers in other states and cities, with lists of alum baking powder companies specific to each location—a massive advertising undertaking.

The cream of tartar war was expensive. The cost was not only in advertising dollars but also in what had happened to the market during the infighting. The alum companies were delighted that the cream of tartar companies were battling one another; it allowed them to gain market share. "The entire south was irrevocably lost to cream of tartar and gained by the alum interests. The west and southwest became enormous users of alum baking powders and alum-phosphate baking powders."[45]

By 1904 cream of tartar was at most 2.5 percent of the total baking powder sold in the South. However, these statistics were deceptive. According to the alum manufacturers, Royal's loss of its cream of tartar business in the South was offset by the alum baking powder it secretly manufactured and sold there. "William Ziegler . . . is to-day the largest manufacturer of alum baking powder in the world, supplying two-thirds of all this great business of the south." However, the difference in price was so great between cream of tartar and alum that Royal wanted to contain alum baking powders in the South. It did not "want these wicked alum baking powders that they are selling in the south to invade the north. That is why they are willing to kill off the people of the south and save the northern people."[46]

In the West and Southwest, baking powder changed foods connected to African American and to Hispanic identity. *What Mrs. Fisher Knows about Old Southern Cooking*, published in San Francisco in 1881, is one of the earliest cookbooks written by an African American. Abby Fisher, born in approximately 1832 in South Carolina to a French father and a slave mother, was an accomplished cook. However, Mrs. Fisher was illiterate, so whoever transcribed the recipes removed almost all of Mrs. Fisher's African American voice, as food historian Andrew Warnes has pointed out.[47]

The first recipe in the book is for "Maryland Beat Biscuit," leavened as usual by manual labor. Fisher's recipes for other breadstuffs and cakes are leavened with yeast (4), eggs (2), baking soda (2), or yeast powder (7). In her Historical Notes to *What Mrs. Fisher Knows*, Karen Hess says "*yeast powder* refers to a mixture of cream of tartar and saleratus, a precursor of baking soda. . . . This historical term . . . was very nearly unique to South Carolina."[48] This is not accurate. "Yeast powder" was Eben Horsford's term for his monocalcium phosphate and baking soda combination. The term "yeast powder" was picked up by other companies and was in use throughout the United States.

Most importantly, it would not have made sense to write a cookbook in San Francisco if one of the main ingredients could not be found there. In fact, a yeast powder skirmish was occurring in San Francisco at the time Mrs. Fisher was writing her book. Rumford had contracted with distributors Church & Clark "to use all proper efforts to secure the introduction and sale of the goods [yeast powder] in California and the Pacific coast generally." Church & Clark would get a 5 percent commission and exclusive rights to the territory. There were some specifics in the contract: the cans of yeast powder were to be "of not less size than Preston & Merrills sold in that market."[49] Two years later the sale of Rumford's yeast powder had met obstacles. Rumford authorized Church & Clark to reduce the price to "at least fifty (50) cents per gross under all competition." Rumford also told Church & Clark to "press the sale with vigor and determination in order to gain the Market."[50]

At the same time, an African American man, Alexander P. Ashbourne, patented a new kind of biscuit cutter. It had a flat bottom surface for rolling out dough. Then the top, which was attached by a hinge, was pushed down on the dough. The top contained nine biscuit cutters in three shapes—circles, hearts, and diamonds—on a spring mechanism so that they cut the dough and sprang back. Ashbourne's ingenious device was like a self-contained work station, because there was also a slot to hold the rolling pin and a hook that fastened the biscuit cutter closed for storing or hanging up.[51]

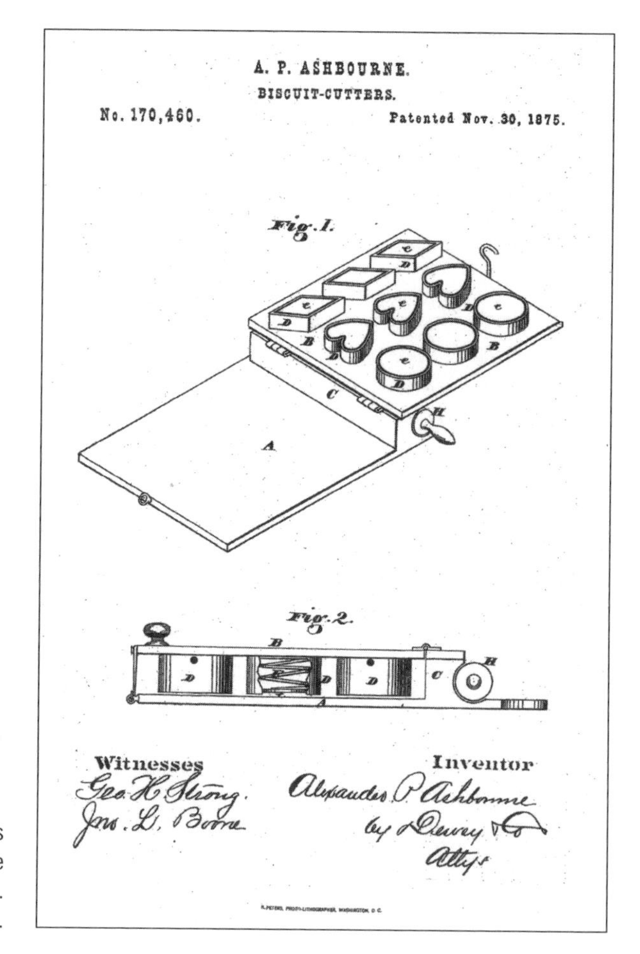

Alexander Ashbourne's ingenious portable biscuit work center, 1875. US Patent Office.

In 1898, also in San Francisco, the "first cookbook written by a Hispanic in the United States" was published.[52] Encarnación Pinedo, an upper-class Latina, was the author of *El cocinero español* (The Spanish Cook) a thousand-recipe cookbook. This was Californio cuisine, the cooking of the Spanish-speaking ranchero elite, the ruling class before the Gold Rush brought California into the union as a free state in 1850. Encarnación's biographer, Victor Valle, points out that she omitted American recipes from her cookbook "as a refusal to acknowledge those who had turned her world upside down."[53]

Although Encarnación does not include American recipes, she does use one important American ingredient: baking powder. It provides the leavening in several different kinds of recipes: semolina rolls; two recipes for cornbread, one from cornmeal, one from fresh corn; two corn casseroles; three different kinds of puddings; and sopapillas, made from fried dough.[54] At that time of transition in cooking, Hispanic recipes, like other recipes, combined traditional ingredients or methods with the modern technology of baking powder. In this case the tradition that was preserved was the metate, the stone that had been used for grinding corn by the Aztecs, which Encarnación used to grind the corn she used to make cornbread leavened with baking powder.

By the end of the nineteenth century, the 387 baking powder companies that Cleveland Baking Powder had listed in its 1888 circular to grocers had grown to 534.[55] The cream of tartar companies had seen their market share—almost the entire market only thirty years earlier—dwindle alarmingly. After the recession of 1893, their high prices had further eroded their market share until by 1900 it was only 20 percent.[56]

With the exception of a few phosphate baking powders like Rumford, almost all of the other 80 percent of the baking powders were alum based. Two of the most important alum-based companies did not even exist until almost the end of the century. Both would become primary combatants against Royal in the baking powder war, and both were in the Midwest.

CHAPTER 6

The Rise of Baking Powder Business

The Midwest, 1880s–1890s

Monopoly must yield to moderation—Impurity must improve or go under—"Calumet" is the standard.

—*Ann Arbor Register*, 1895

"Clabber Baking Powder" A NEW INVENTION, considered the healthiest baking powder on earth. Where once introduced, it takes the lead over anything else.

—Clabber Baking Powder advertisement, Terre Haute, Indiana, 1900

In 1893 Chicago announced to the world that the American Midwest, as a region by itself, had arrived as a player on the world stage. From May to October more than twenty-one million people visited the World's Columbian Exposition to celebrate the four hundredth anniversary of Columbus's arrival in the western hemisphere.[1] The fair left lasting imprints on American culture with the creation of the Pledge of Allegiance and the celebration of Columbus Day as a national holiday. The Columbian Exposition also marked the debut of one of the most famous icons in advertising history, and it was connected to baking powder: Aunt Jemima.

In 1889, in St. Joseph, Missouri, two men who knew nothing about food, chemistry, or cooking bought a bankrupt flour mill and thought they could rescue it by creating a specialty product. Chris Rutt and Charles Underwood combined flour with phosphate of lime and baking soda—in other words, baking powder. The cook only had to add milk.[2] They experimented in Rutt's kitchen until they got the proportions right, and then they proudly presented their creation: "Self-Rising Pancake Flour," the first in the world. The world was not impressed. Renaming the mix after the popular minstrel show character Aunt Jemima did not improve sales. In January 1890 Rutt and Underwood sold the business to R. T. Davis, the owner of a flour mill. Davis added another ingredient to the formula: powdered milk.[3]

Powdered milk was new. The process for drying milk had been invented in 1866 by a German-born Swiss, Heinrich Nestlé. Appalled by the 20 percent infant mortality rate in Switzerland, Nestlé wanted to find a way to preserve milk for babies whose mothers could not nurse them. Before refrigeration, even in cold climates, milk was highly perishable and could spoil easily. Nestlé experimented in his kitchen until he produced dried milk that was as "'fresh and wholesome' as Swiss milk 'straight from the cow's udder.'"[4] The addition of powdered milk made the pancake mix not only consistent but convenient as well, because the cook only had to add water.

Davis also came up with one of the most brilliant marketing ploys in history. He decided to capitalize on the name of the pancake mix. Davis connected the pancake mix to popular racist propaganda about the plantation South and the myth of the happy slave. He hired a real former slave, Nancy Green, to play the fictitious slave Aunt Jemima. At the Columbian Exposition in Chicago, inside a booth shaped like a giant flour barrel, Aunt Jemima came to life. She walked, she talked, she sang, she cooked for fairgoers sporting pins with a picture of the smiling Jemima and the words "I'se in town, honey." Merchants put in more than fifty thousand orders for Aunt Jemima pancake mix.[5]

Also at the Columbian Exposition, historian Frederick Jackson Turner presented his thesis that the American West was "closed"—settled, based on population density in the 1890 census. As the real West became tame, the mythic West, especially the Plains West, became diffused throughout American culture. Adjacent to the fairgrounds, real-life army scout Buffalo Bill brought the West of legend to the city. His show featured real Western celebrities like Wild Bill Hickok and Native American Sitting Bull, along with Ohio native sharpshooter Annie Oakley. Buffalo Bill's Wild West Show reenacted Custer's Last Stand, Native Americans attacking settlers, and the Pony Express. America's fascination with the West continued in dime novels, on stage, in films, and on television into the twenty-first century. It was also memorialized in a new Midwestern baking powder.

Chicago, Illinois: Calumet Baking Powder

In 1888, the year that Ziegler and the Hoaglands were embroiled in their lawsuit for control of Royal in New York, a top salesman for Dr. Price's baking powder jumped ship and became a competitor. Ohio native William Monroe Wright (b. 1851) decided it was time to go into business for himself—in Chicago, Dr. Price's headquarters. Evidently, Wright did not have a contract with Dr. Price, or if he did, it did not contain a noncompetition clause like the one that other baking powder companies used. In 1891 a contract between Rumford and a salesman in Oregon stated that the salesman was "not to sell or be interested in the manufacture or sale of any goods which can be used for the same purpose as Horsford's baking powder."[6]

Wright did not want to make a cream of tartar baking powder like the one he had been peddling for Dr. Price. He wanted to create something more modern. He scraped together thirty-five hundred dollars and partnered with a chemist. George Campbell Rew (b. 1869), a Chicago native educated at the University of Michigan, graduated from the University of Illinois School of Pharmacy in 1889.[7] The baking powder Wright and Rew created contained baking soda and a cornstarch buffer, like other baking powders. However, the leavening agent was sodium aluminum sulfate. Wright and Rew also added albumen—powdered egg white—which has a dramatic foaming effect when mixed with liquid. This would prove particularly useful in demonstrations to housewives. It would also cause trouble with the press, competitors, and federal agencies.

Wright called his baking powder Calumet, which means "peace pipe." Calumet is also a common place-name throughout the Midwest. It was a Menominee village on the east side of Lake Winnebago. Other places with the name Calumet are Calumet County, Wisconsin, incorporated in 1836;[8] Calumet Township in Michigan, a company town named after the Calumet Mining Company in 1866;[9] and Calumet, Iowa, founded in 1887.[10] Calumet City, Illinois, did not exist until 1893, when West Hammond changed its name.[11] In addition to having a name with immediate brand recognition, Wright also had a younger brother who had died; his name was Calumet.[12]

The Calumet baking powder can label was bright orange and featured the solemn face of a Plains Indian warrior chief in full eagle-feather headdress. His long black hair is in two braids that come down in front of his shoulders. Visually arresting, the label also capitalized on urban America's fascination with the West and racist mythology about Native Americans.

Wright's use of a distinctive picture, color, and name for his product was part of the "explosion of branded consumer goods introduced to the mass market" in the 1880s, along with Campbell's soup, Coca-Cola, and Heinz.[13] Wright intended to create a national brand from the beginning, even though the business was based in his small home kitchen, where he and his family put the powder into cans and then put labels on them. Name brands, unlike generic anonymous foods sold in bulk bins, offered consumers reliability and consistency. In 1890 Wright began to send advertising materials to grocers throughout the United States. He touted Calumet as "the acme of perfection, economy, purity, and strength."[14]

A great part of the appeal of the new alum baking powders like Calumet was their price. A Calumet advertisement in the *Ann Arbor (MI) Register* in 1895 attacked cream of tartar baking powders directly, on the grounds of price and monopoly:

> In 1872 Cream-of-Tartar was 40 cents per pound; in 1892 it was 19 cents per pound. Have the high-price Baking Powder monopolists reduced their price? They have not!

> "Calumet" was the first, and is the only high grade Baking Powder offered to the public at a moderate price. Its motto is:—
>
> Monopoly must yield to moderation—Impurity must improve or go under—"Calumet" is the standard.[15]

However, Calumet was not the only new alum baking powder in the Midwest.

Terre Haute, Indiana: Crossroads of America

From the beginning, Terre Haute has been a study in diversity. It is a town with a French name—Terre Haute (high ground); in a county named after an Italian explorer—Vigo; in a river valley named by Native Americans—Wabash (pure white); in a state created by the descendants of Englishmen—Indiana.[16] "Crossroads of America," says a historic marker at the corner of Wabash Avenue and Seventh Street in Terre Haute, Indiana. Wabash Avenue, now U.S. Highway 40, began as the east-west National Road, which reached Indiana in 1829. It connected Terre Haute with Baltimore, Maryland, its eastern terminus; and Vandalia, Illinois, its western terminus. Its intersection with Seventh Street, the road that became north-south U.S. Highway 41, forms the Crossroads of America.[17]

Admitted to the Union in 1816, Indiana was attractive to migrants for many reasons. Title to land was clear, because it had been surveyed by the federal government under the Land Ordinance of 1785 and because the Native American population was on reservations by 1821.[18] The Northwest Ordinance in 1787 declared that America's new Northwest Territory, which included Indiana, would never be open to slavery, so small family farms could prosper because they would not have to compete with slave labor.

Transportation and fresh water were readily available to Terre Haute from the Wabash River, seven blocks west of the Crossroads of America. Just south of Terre Haute, the Wabash flows into the Ohio River. East and upstream on the Ohio River are Louisville, Kentucky; Cincinnati, Ohio; Wheeling, West Virginia; and Pittsburgh, Pennsylvania. Downstream, the Ohio River feeds into the Mississippi.

Canals also connected Terre Haute with ports thousands of miles away. In London, the center of world banking, the Rothschilds bought the bonds; in Indiana, the Irish dug the ditches.[19] One canal went north from Terre Haute to Lake Erie, then along the Erie Canal to the Hudson River and New York City. The importance of the Erie Canal to the Midwest went beyond transportation and communication. Before the canal was completed in 1825, Midwestern trade was primarily downriver to New Orleans. If that had continued, the Midwest would have been bound to the Confederate South economically and politically. Instead, the Erie Canal caused a massive economic and political shift to the Northeast.[20] The Midwestern states provided raw materials and agricultural products to New York, colonies that fed the Empire State. After 1850 the Chesapeake and Ohio Canal connected to the

National Road, which meant transportation straight through from Terre Haute to Washington, DC.[21]

Two blocks north of the Crossroads of America are train tracks. They are audible and visible from the house where Eugene Debs, president of the American Railway Union, grew up. By the early 1850s, Terre Haute was crisscrossed by multiple rail lines. The Terre Haute and Richmond Railroad went to the east, the Terre Haute and Crawfordsville to the south, the Terre Haute and Alton to the west.

Newcomers to Indiana settled into roughly three different cultural, political, and culinary sections. The northern section was settled by Northerners, New England Yankee abolitionists who farmed wheat, dairy, and fruit. Southern Indiana, settled by Scots-Irish slave-owning sympathizers from the Appalachian foothills, was corn and hog territory. The middle section, where Terre Haute is located, was diverse: Quakers, Dutch, Pennsylvania Dutch, and free blacks. In the 1840s and 1850s, Irish immigrants fleeing the potato famine and German immigrants fleeing the failed democratic revolutions also settled in the middle zone.[22] These settlers were part of America's Great Migration. As historian Daniel Walker Howe points out, "Never again did so large a portion of the nation live in new settlements."[23]

The migrations and population expansion into new territories created new opportunities for businesses. Terre Haute also benefited from the transportation revolution that occurred in Jacksonian America. At the intersections of rivers, canals, roads, and railroads, Terre Haute was ideally situated for business.

It was exactly these advantages about Terre Haute that appealed to Francis Hulman. The twenty-eight-year-old German Catholic bookkeeper had arrived in Cincinnati in the 1840s in response to a letter from his older brother, Diedrich.[24] On November 10, 1849, Francis repeated the pattern and wrote to his younger half-brother, Herman, in Lingen-En-Ems, in the province of Hanover, Germany, near the border with the Netherlands. Francis was planning to move west and change professions. He did not like the "fancy business" of importing "toys, jewelry, and toilet articles" that he and his partner were in.[25] Francis wanted something more substantial and profitable, and not seasonal. He also felt that his partner was too conservative and not willing to take the business risks necessary to capitalize on the opportunities Francis saw everywhere around him. So he decided to move three hundred miles west to Terre Haute, get a new partner, and go into the wholesale grocery business. Herman would be able to help him.

Herman had been in the grocery trade since the age of fifteen, when he began a three-year apprenticeship to a wholesale grocer. In the European system the apprenticeship cost Herman $100 per year, a hefty sum.[26] In his letter, Francis asked Herman to send the $350 he needed to go into business, and also to come to the United States himself:

> We have chosen a place called TERRA HAUTE [*sic*], three hundred miles west of here in the State of Indiana. This is a place of about 6,000 inhabitants, very well

> laid out, flourishing and growing rapidly, in a clean and healthy location. Besides this, the territory is rich with wealthy peasant farmers in the surrounding country and beautiful spreading meadows. This place is, as I have said, one of the best in this vicinity and if we have any luck we are sure to make money.[27]

Francis also appealed to Herman by contrasting a life of freedom, opportunity, and dignity in America with a rigid, limited life in Germany:

> O Herman! Herman! follow my advice. There is still time! You will lead an entirely different life, be a different person, a free man, independent, and a republican who is conscious of his worth and dignity as a man. In this free and happy America, poverty and ignorance do not reign, one can express his opinion freely, and there is no censorship. The laws are good and wealth and well-being reign everywhere.
>
> . . . You will have quite a different life. . . . You will keep one or two horses; you can go horseback riding once or twice a week and hunting or fishing at least every Sunday. You will eat and drink whatever you wish, with no strong master and no fury to embitter your life. You will be free and independent, your own master. You can act freely in every respect as it becomes a man.[28]

Francis was a salesman, and he was selling America to Herman. But Francis was also a businessman. He was not about to give anything away, even to relatives, and not even to relatives who helped him: "When you have been here several years and have become acquainted with the customs, the language, etc., we will advance you credit. You can then start your own business, alone or with a partner and you will be sure to make money. Instead of poverty, misery, and disagreeableness, you will spend your life in happiness, well-being, rest, and contentment. What a difference!" What Francis proposed to Herman was a quid pro quo arrangement: you loan me money now, I will loan you money later. He also presented Herman's $350 as an investment with unlimited future dividends.

Francis Hulman's letter described antebellum Indiana, its Golden Age. In the 1850s Indiana was post-frontier and preindustrial. In rural areas the older 50 percent of the homes built of log were giving way to new houses of brick or frame. The recent forced removal of Native Americans had opened the land to white farmers. Land was plentiful. Forests of oak, beech, maple, hickory, ash, walnut, poplar, elm, sycamore, and cherry provided not only lumber but also animals to hunt and trap. The average farmer cultivated sixty-five acres and had a wife, four children, and no mortgage. When he sent his surplus wheat or hogs to market, competition among canal, river, railroad, and the National Road made transportation affordable. The principal rural occupations were flour milling, saw milling, and hog packing.[29] A farmer writing in 1859 explained why he chose Indiana: "I would go west but Iowa is too cold Kansas not settled and Missouri a slave state. Indiana is good enough for me."[30]

The farmer's counterpart in Indiana's small towns owned a business capitalized at $3,500 and with four employees. The principal occupations in town were making shoes, furniture, liquor, and machinery. There were no wide disparities in income. Like the other states in the Old Northwest, land values doubled in the decade, from $10.66 per acre in 1850 to $21.16 in 1860.[31] Optimism was in the air.

On March 12, 1853, Francis Hulman dissolved his partnership and went into the wholesale and retail grocery business for himself in Terre Haute.[32] In 1854 twenty-three-year-old Herman Hulman left Germany and joined Francis in the grocery business. They both began writing persuasive letters and sending money to their younger brother, Theodore, who was still in Germany. In 1857 seventeen-year-old Theodore arrived to join them.[33]

The Hulmans were just three among the seven million Germans who immigrated to the United States, the majority to the Midwest.[34] They settled primarily in the upper Ohio, Missouri, and Mississippi river valleys. Germans are a classic example of a "pull" migration: they came to the United States because opportunity beckoned. They were not pushed out of their home countries because their lives were in danger from famine or for religious or political reasons. Unlike the Irish, Italians, and other ethnic groups that had little money and few skills, many of the Germans arrived with some money, education, and professions. They did not have to stay in the East Coast cities where they arrived. They could afford to move inland and to buy land. Clusters of German immigrants created Little Germanies in New York, Chicago, Milwaukee, St. Louis, and Cincinnati.[35]

The 1850s, when Herman and Theodore came to the United States, was a peak immigration period for Germans. In 1854 Herman was one of the 215,000 Germans who came to the United States. By 1860 approximately 1.3 million Germans lived in the United States and supported two hundred German-language publications.[36]

A traveler in Indiana in the 1850s said, "A backwood farm produces everything . . . except coffee and rice, salt and spices."[37] The farmer also did not produce dry goods, shoes, or farm equipment. Those were the items the Hulmans carried in their store.

By 1858 the store was thriving under the management of the three brothers, who were very close. Unaffected by the Panic of 1857, the business had moved to larger quarters, a three-story building plus basement, on the corner of Terre Haute's Main and Fifth streets.[38] Francis went to Germany to visit his mother and family for two to three months. On the trip back to the United States, there was an explosion on the ship. Francis, his wife, and their three-year-old daughter were among the more than 120 passengers who died. A grieving Herman, now twenty-seven, was thrust into managing the business.[39]

The store continued to do well. The Hulmans began to sell national brands. Among them was Royal Baking Powder, first mentioned in company records in May 1869.[40] Other national brands included Van Kamp's; Procter & Gamble; Libby,

Table 6-1. Prices in the Hulmans' Store, 1853

Beans	.25	5 pounds
Beef	.04	Pound
Castor Oil	.75	Bottle
Cheese	2.78	24 pounds
Coffee	21.04	165 pounds
Flour	6.75	Barrel
Peach Brandy	16.75	16¾ gallons
Salt	4.00	Bag
Smoking Tobacco	3.60	Bag
Sugar	.25	50 pounds
Whiskey	.50	Gallon
Wine	.50	Bottle

Source: A. R. Markle and Gloria M. Collins, *The House of Hulman: A Century of Service, 1850–1950* (Terre Haute, IN: Clabber Girl, 1952).

McNeill & Libby; Lipton; and Lorillard.[41] The Hulman brothers diversified. They bought a small distillery and then used the leftover mash to feed cattle and pigs that other businessmen sent to them.[42] By 1878 their distillery was the largest in the United States, consuming forty-four hundred bushels of corn each day. The grocery, too, was one of the largest in the country.[43]

The Hulmans exemplified the nineteenth-century boosterism that was widespread in the United States as not only businesses but cities as well engaged in competition for markets. The Hulmans were keenly aware of the importance of the technological innovations appearing in America, both for Terre Haute and for their businesses. A symbiotic relationship grew between businesses and the city, as each aided in the growth of the other. If there was a committee to build, invent, or explore something for the benefit of Terre Haute, a member of the Hulman family was on it, often as the founder or president. The Hulman family also invested heavily in infrastructure and new technology. The businesses they supported were technologies introduced in the nineteenth century: railroads, telegraph, electric lights, fire alarms, water, sewage, gas, drawbridges, iron and steel companies, and lumber and paper mills.[44]

The technology infrastructure was important to the Hulman businesses. On October 20, 1877, telephones connected the Hulman distillery and the wholesale house. The telephone was new technology, exhibited for the first time at the Centennial Exhibition in Philadelphia only the year before. The city of Terre Haute did not get a telephone exchange until 1880, three years after the Hulmans installed telephones in their business.

All members of the Hulman family were extraordinarily gifted and high achievers both inside and outside of business, including the children. The men competed in sports and animal breeding; the women in cooking, sewing, and canning. The men competed in two new sports, bicycle racing and baseball, and won trophies at the local, state, and national levels. On his farm on North Thirteenth Street, Diedrich experimented to improve the quality of fruits and vegetables. He also imported queen bees from Egypt and Italy and was the leading bee expert in Indiana. Everyone in the family entered contests in the Vigo County Fair. Diedrich and his son Otto usually finished first and second with horses and cattle—until 1877, when Herman entered eleven steers and won. They also raised various breeds of pigeons and chickens and entered contests for honey, fruits, and vegetables.[45]

The women and girls won prizes, too. Katie Hulman started winning prizes when she was nine. Her first was for quilting; she won others for embroidery and crocheting.[46] Diedrich's daughter Sophie won a prize for a jar of honey. In later fairs she entered fruits and vegetables that she raised herself. Wives Loriette (Diedrich's wife), Antonia (Herman's wife), and Sophia (Theodore's wife) were accomplished at all types of needlework. They won prizes for pleated shirts, trousers, embroidery, petit point, crochet, quilts, and rugs. Their baked goods—"fancy breads, rolls, cakes, and pastries"—were prizewinners, too. So were their canned fruits and vegetables, pickles, fruit and dairy butters, jams, and jellies.[47]

In 1879 Herman decided to concentrate on the grocery and sold his ownership in the distillery.[48] Great changes came to the Hulman family in the mid-1880s. On April 17, 1883, Herman's wife, Antonia, died. He never remarried. No matter how busy he was, every day until the end of his life he made a pilgrimage to his wife's grave.[49] On May 2, 1885, Herman restructured the firm to bring in his son Anton as a partner, along with Benjamin Cox. Herman and Cox had been involved in various business capacities since 1869, when their separate grocery firms had merged.[50] Through the 1880s and 1890s, the Hulmans weathered explosions and fires at the distillery as well as being fined by the Internal Revenue Service for taxes and for operating the business on Sundays.[51]

By 1893 the Hulmans were one of the leading families in Terre Haute. They celebrated their fiftieth year in business with the opening of a huge new seven-story building that took up a full block at Ninth Street and Wabash Avenue. At the rear of the new building was a five-story spice mill. On the third floor, coffee was blended, roasted, and ground. On the second floor, three people prepared baking powder and put it into cans.[52] The following year, Hulman & Co. purchased a printing press so that they could control printing of catalogues as well as labels for the cans. It was faster, easier, and cheaper than outsourcing printing.[53]

On September 28, 1893, Hulman & Co. invited five thousand people from all over the United States, including journalists, for the grand opening of the new building. Guests arrived on special trains. They were treated to an afternoon at the fairgrounds to watch trotting horse races, to attend the largest banquet in Indiana (held at the Terre Haute House Hotel at the corner of Wabash and Seventh Street), and to view almost an hour of fireworks.

The mayor of Terre Haute and other local dignitaries spoke. The final speaker was Terre Haute native Eugene V. Debs, former city clerk (1879–1883) and member of the Indiana legislature (1885).[54] More importantly, just three months earlier, on June 20, 1893, Debs had organized the "first industrial union in the United States," the American Railway Union (ARU) in Chicago.[55] A few months after the banquet, in the spring of 1894, Debs led the ARU in a successful strike against the Great Northern Railroad. The deep affection and high regard that Debs, a champion and organizer of labor, had for Herman Hulman, a businessman, spoke volumes about Hulman's character.

The two men had an extraordinary relationship. Debs was the son of Germans who, like Herman, had emigrated to the United States after the failed democratic revolutions.[56] Herman and Debs had been friends since 1874, when Debs had worked for Hulman as a billing clerk.[57] Debs, born in 1855, was twenty-four years younger than Herman, young enough to be his son. At the store opening in October 1893, Debs spoke at length about Herman's character, integrity, and philanthropy: "The maimed, the infirm and the friendless will hold his name in consecrated remembrance long after he has been gathered to his fathers."

Debs also spoke about Herman's workers, the traveling salesmen and drummers, as a force for civilization in "the cabin home on the frontier." Debs equated civilization with material goods that "redeemed [the settlement] from old time customs and costumes" and said that through new cooking stoves and newspapers "a revolution is inaugurated." Overall, "the world is made brighter and more beautiful by the work of the commercial traveler." In closing, Debs spoke again of Herman: "With him integrity was not an acquired virtue but an inherent governing force which he could no more have disregarded than he could have regulated the throbbing of his heart. His name stands for all things of good report, and will be forever associated with the progress of the Prairie City."[58]

On October 13, 1893, Herman handwrote a heartfelt thank-you note to "My dear friend Debs":

> I assure you no congratulation has been as appreciated like this one coming from you. I know that comes from the soul and means what it says. I wish to thank you with my whole heart for your kind wishes and assure you that this feeling of esteem & friendship is mutual and therefore appreciated your expressions so much more than coming from ordinary acquaintances.
>
> True friendship is not often met with and it seems to me such is not cut apart by Death even, and if so, we will meet hereafter and enjoy our company in a better life—where no parting is.—
>
> This is the sincere wish of your friend,
>
> H. Hulman[59]

In 1900 the population of Terre Haute was almost thirty-seven thousand, and Hulman & Co. was worth almost one million dollars.[60] In half a century the Hulmans had become part of the influential business class that ruled Terre Haute. Part of the Hulmans' prosperity was because they innovated, constantly seeking new products and new businesses. One of the new businesses that caught Herman's attention was baking powder. He had begun manufacturing it in 1879. His first two baking powders, Crystal and Dauntless, were sold in Vigo, Sullivan, Parke, and Putnam counties in Indiana. "Dauntless" was also the label for Hulman & Co.'s proprietary coffee blend. Herman bought the coffee beans green and then roasted, blended, and ground them onsite in a spice and coffee mill behind the store.[61] He kept experimenting with and improving on the names and formulas of his coffee and baking powder, and in 1887 he introduced a baking powder called "Milk." In the 1890s Herman became intent on creating one baking powder—one formula for a quicker acting, better baking powder. He hired chemist Gus Gorrell to do the job. In secret and at great expense they experimented for years.[62]

Baking powder was in widespread use in the German and German American communities by this time. Henriette Davidis, a German woman, wrote *Praktisches Kochbuch* (*Practical Cookbook*), in Germany in 1847. It became a best seller that went to twenty-one editions and extended after her death on April 3, 1876. The first American edition, in German, was published in 1879, in Milwaukee, Wisconsin. The publisher wanted to capitalize on the spirit of the American centennial in

Herman Hulman. Courtesy of the Vigo County Public Library Community Archives.

1876, so its theme was bicultural: the wholesomeness of German food, along with the foods of "our American homeland." It included recipes for American baked goods made with baking powder: baking powder biscuits, cookies, cornbread, and cakes.[63]

The first English translation of Davidis's book, in 1897, includes a section at the end titled "The American Kitchen." These fifty-three pages contain recipes for "various dishes prepared in styles peculiar to cooking as done in the United States": clam chowder, succotash, baked beans, tomato catsup, grape jelly, and others.[64] This section also includes thirty-nine recipes for "Cakes, Cookies, etc." Of these recipes, baking powder leavens twenty-seven; baking soda, nine; and cream of tartar plus soda, one. Baking powder also appears in other foods: apple fritters, strawberry shortcake, and buckwheat griddle cakes.[65]

However, it is the German recipes in the body proper of the cookbook that show how deeply baking powder had penetrated German American cuisine and culture by the end of the nineteenth century. For Germans in America, bread remained yeast-risen, but baking powder biscuits had become the default biscuit. Labeled simply "Biscuits No. 2," it was unnecessary to include "baking powder" in the name. German cooks added baking powder to Lydia Maria Child's 1832 cup cake ("1-2-3-4 Cake") recipe. In some recipes—for example, "Portuguese Coffee Cake"—ancient Middle Eastern flavorings such as orange flower water are side by side with the new technology of baking powder.[66]

Baking powder altered traditional desserts used in ritual celebrations and connected to German identity. These included "Speculaci," or "Tea Tarts for the Christmas Tree"; "Hohenzollern Cakes," named after the ruling family of the Austrian Empire; *pfeffernüesse* ("White Rifle Nuts"); and honey cakes.[67] It is impossible to tell exactly how many recipes used baking powder, because the instructions at the beginning of the cake section are: "The addition of baking powder in biscuits, bread and almond cakes and the like will tend to make them lighter."[68]

Cookbooks like Davidis's connected German-speaking women in Europe and the United States and aided in the transnational and cross-cultural transmission of knowledge about baking powder. No doubt the ready acceptance and widespread use of baking powder by German American women influenced European German women to use it. Women like those in the Hulman family, who routinely won prizes for their baking, preserves, and needlework at county fairs, would have corresponded with mothers, sisters, and aunts in Germany about this exceptional new way to leaven cakes and leave no odor, the way that ammonia could if it was not used properly. When baking powder did become commercially available in Germany, German housewives used it the same way that German American housewives used it: for dessert, not bread.

By 1893 baking powder was accepted and in use enough in Germany that manufacturing it finally became profitable. Dr. August Oetker, a pharmacist, sold a baking powder under his name. The formula was the one that Horsford had given to Liebig in the 1860s. Oetker bought it in 1890 from Luis Marquart, the son of the owner of one of the baking powder companies that had failed. Oetker advertised his baking powder for use in desserts, not to replace the yeast in bread.[69] In the twenty-first century, Dr. Oetker is still a brand of phosphate baking powder, made from sodium acid pyrophosphate, sodium bicarbonate (baking soda), and cornstarch.[70]

The End of the Century

The year 1893 saw major changes at Rumford. A severe recession began that would reduce Rumford's dividends to a dribble. The directors discontinued the bonus to the employees that had been an annual gift since 1886.[71] The most severe blow to the company was that Eben Horsford died suddenly from a heart attack on January 1, 1893.[72] His memorial service was held at Wellesley College. "Wellesley has known no friend—with the shining exception of her first two friends and founders—so liberal as Professor Horsford; and his name will be forever honored in her gates. He has richly endowed her library; he has provided a fund for scientific apparatus; he has established, for thirteen professional chairs and for the presidency, the grant of the Sabbatical year, with a system of pensions for retired officers."[73]

Control of Rumford remained in Horsford's family but with treasurer Newton Arnold at the helm. In 1898 Rumford tried to strengthen its position in Chicago and the Midwest by forming an alliance with the Woman's Christian Temperance Union (WCTU). Rumford would pay the WCTU "at the rate of five (5) cents per pound of powder represented by the labels" if the WCTU used "every reasonable and proper effort . . . to increase the use and sale of Rumford Baking Powder." At the time, Rumford baking powder was selling for thirty cents per pound. This was a way to "help the local Union in every city and in every town to help themselves. Instead of asking for money, we will put money into their treasury."[74]

In Chicago in 1899, William Wright elevated his son to secretary of Calumet. Twenty-four-year-old Warren Wright, born September 25, 1875, educated in public schools and at Bryant and Stratton Business College in Chicago, had been working at the company for nine years.[75] Restless, he had dropped out of high school at the age of fifteen to become a cowhand on a Texas ranch and then a bill collector for a grocery before he joined Calumet.[76]

Also in 1899, Herman Hulman's years of experimentation to create his perfect baking powder finally bore fruit. He was satisfied with his new, sodium aluminum sulfate (alum)–based baking powder. Herman set aside part of the spice mill to manufacture the baking powder in 330-pound batches through a series of laborious, disjointed processes. After the ingredients were mixed, they were offloaded into a wooden wagon, dumped into large cans, then scooped into smaller cans, labeled, and put into cases of twenty-four cans each. Most of the work after the mixing, which was secret, was done by seven women who worked an eleven-hour day, 7:00 a.m. to 6:00 p.m.[77]

Herman called the baking powder "Clabber," after a Midwestern term for delicious, thick, fermented milk, which housewives used as a natural leavener. The label reflected the connection to dairy: in the foreground, a young woman worked a butter churn.[78] Herman was ready to put his new Clabber baking powder into distribution. His timing could not have been worse.

William Ziegler had finally decided on a strategy to annihilate his alum competitors. He did not want a close vote to outlaw his competition; he wanted a guarantee. The first state to outlaw alum baking powders would have to be carefully chosen. That state would have to be in the Midwest heartland, the headquarters and stronghold of the alum baking powders. That state would have to be willing to provide a virtuous example of pure food to the rest of the country. That state would have to have legislators eager to be on the cutting edge of science and to set bold new public policy. And who could easily be persuaded by cash.

Ziegler went to Missouri.

CHAPTER 7

The Pure Food War

Outlaws in Missouri, 1899–1906

In the early [eighteen] eighties the Lobby grew into a regular and systematic business.

—"The Great American Lobby," *Leslie's Popular Monthly*, 1903

Big businesses contributed to all campaign funds, and this is the first step toward corruption everywhere. It is wholesale bribery.

—Lincoln Steffens, "Enemies of the Republic," *McClure's Magazine*, 1904

No pure food legislation can get through Congress without my consent.

—Daniel J. Kelley, lobbyist, Royal Baking Powder, *Chicago Royal Herald*, 1903

Missouri was known as a "robber state."[1] Fifteen years of Jesse James withdrawing other people's money at gunpoint from banks throughout the Midwest and then finding sanctuary in Missouri had earned the state that label. James and his gang had had public sympathy during Reconstruction when they continued their Civil War bushwhacking tactics and committed daring new crimes: robbing banks during daylight and trains while they were moving. But in the booming Gilded Age economy, Jesse James was bad for business. In 1881 Governor Thomas Crittenden put a bounty of ten thousand dollars each on Jesse and his brother, Frank. The railroads put up the money.[2] When members of the James gang approached the governor to say that Jesse could not be taken alive, the governor reminded them that he had the power to pardon. On April 3, 1882, Bob Ford shot Jesse James in the back. Ford was convicted, sentenced to hang, pardoned, and rewarded.[3] Having the state's chief executive solicit murder did not help Missouri's reputation.

By 1899 lobbyists had enjoyed a corrupt relationship with government in Missouri going back to at least 1878, when Colonel William H. Phelps, "The King of the Lobbyists," became the railroad representative in the Missouri legislature. Phelps, a lawyer who had arrived in Missouri by stage coach, contributed to Missouri's reputation as an outlaw haven. He had survived being shot twice in open court by an opposition attorney, and had returned the favor fatally with a double-barreled shot gun in a high-noon ambush when his opponent was unarmed. After a trial and an acquittal, Phelps was back in business. As the representative of the Gould railroads, Phelps was not standing in the lobby of the capitol at Jefferson City begging for favors. He was sitting behind a curtain in the back of the senate chamber, sending runners out with written instructions ordering bought-and-paid-for senators how to vote. The people of Missouri had traded a masked bandit for a curtained one; crude robbery for corporate. The "Show Me" state had become the "Show Me the Money" state.

From 1899 to 1906 the baking powder war was waged on two fronts: state and federal. While baking powder companies were fighting for control of state legislatures, they were also battling for supremacy at the federal level, in both houses of the U.S. Congress. The war was also between two new entities in America: the lobby and the national press. Lobbying—businesses giving gifts to influence politicians—began with the railroad in the 1880s when the states started to pass legislation to control the railroads. The practice soon spread to other businesses and included outright bribery. The national press was a new kind of journalism headquartered in New York City involving monthly magazines with articles of countrywide interest instead of daily papers with local news.

On May 11, 1899, a little more than two months after the Royal Baking Powder Trust was formed, the Missouri state legislature passed a one-sentence law: "That it shall be unlawful for any person or corporation doing business in this state to manufacture, sell or offer to sell any article, compound or preparation, for the purpose of being used or which is intended to be used in the preparation of food, in which article, compound or preparation, there is any arsenic, calomel, bismuth, ammonia or alum."[4] On its face, this was a simple, run-of-the-mill, pure food law intended to protect the residents of Missouri from adulterated food. It looked like enlightened public policy. On closer inspection, however, it was a law with the hidden agenda of private gain masquerading as public good. The crux of the law was the last word—"alum"—designed to shut down Royal's competitors. Royal also targeted ammonia, which it had recently removed from its own baking powder. Stopping the use of ammonia in food would knock any remaining baking powder companies that still used it out of business. It would also block the sale of hartshorn, used in Germany and by German Americans.

The Missouri law did what Royal intended: it unleashed the coercive powers of the state on its own citizens on behalf of private business, giving Royal a personal police force for free. When the law went into effect on August 22, 1899, many small companies that were manufacturing alum baking powder went out of business instantly. It also forced alum baking powders off the shelves. The companies that had manufactured those baking powders lost not only the profits that would have come from selling the product, they also lost the product, one million dollars' worth.[5] Twenty shopkeepers learned about the law for the first time when they were jailed for selling alum baking powder.[6]

The anti-alum law crushed small baking powder manufacturers in Missouri, but it galvanized the large manufacturers in the Midwest. From October 26 to 28, 1899, thirteen baking powder manufacturers and four chemical companies met in New York City, Royal's home, and created the American Baking Powder Association (ABPA).[7] The motive forces behind the association were two Chicago-based baking powder companies, Calumet and Jaques, which produced KC baking powder. The officers and executive committee of the association were the major Midwestern manufacturers of alum baking powder from Chicago; Dayton, Ohio; St. Paul, Minnesota; and East St. Louis, Illinois. They paid annual dues of fifty dollars for an "A" membership if they had sales of more than ten thousand dollars per year, or twenty-five dollars for a "B" membership if they had sales less than ten thousand dollars. They also paid ten cents per one hundred pounds of alum they purchased each month.[8] They urged other manufacturers of alum baking powder "to join this body for mutual protection against the most scandalous misuse of corporate power in modern times."[9] In 1900 the association's sixty-one member companies produced seventy-five million pounds of baking powder per year, approximately three-fourths of the alum baking powder in the United States.[10]

The association's secretary-treasurer was Abraham Cressy Morrison, who devoted his life to popularizing science, especially chemistry. Morrison stated the purpose of the association: "our desire to freely transact the business on which our livelihood depends, and our desire to stand before the community in the light of honorable businessmen." These men, after all, had been accused of poisoning Americans. They realized that they were engaged in a "life and death struggle," because Royal's "very existence depends on the destruction of our business."[11]

The ABPA decided that although a lawsuit against Royal for the blind advertisements was feasible because they were libelous in the aggregate, it would take too long and would not reach consumers directly. The most productive strategy would be to fight Royal on its own terms: by lobbying and advertising. The association launched a massive campaign to influence legislators and newspapers. They hired eighteen or twenty typists for several weeks and wrote more than twenty thousand letters to state legislators—three letters each—informing them of the Missouri bill and that Royal either already had introduced or would be introducing it in their legislature. The association also sent three thousand letters to newspapers across the country, warning them about Royal's advertising tactics.[12]

Regarding consumers, the ABPA realized that it had a massive public relations problem on its hands. In the age when corporate public relations were shifting from robber baron William Vanderbilt's 1882 "The public be damned!" to politician William Gibbs McAdoo's 1908 "The public be pleased," the alum baking powder manufacturers had neglected to inform the public about their product.[13] Instead, they had passively relied on the fact that people used their product "*in constantly increasing quantities with no deleterious effect*." They berated themselves: Royal would not have been able to gain so much ground if the alum manufacturers "had earnestly advertised the wholesomeness of their product and had as boldly in those early days announced the superiority and greater economy in the use of alum" and also informed consumers that alum baking powders were "in harmony with the scientific progress."[14] The association was also aware of what they had that appealed to consumers: "our powder [is] more effective and much cheaper than theirs."[15]

For six years, until passage of the Pure Food and Drug Act in 1906, Royal and the American Baking Powder Association fought a war for control of the legislature of almost every state and territory. Royal introduced the exact wording of the Missouri bill in other state legislatures year after year, sometimes two or three times in one session. Sometimes the anti-alum wording was added as a rider after a bill had been approved in committee, or tacked on as a last-minute addendum. The association was alert to this and parried Royal's every thrust. In spite of persistent, tenacious attempts by Royal year after year, no other state passed an anti-alum baking powder law.

Indiana was anomalous as far as baking powder legislation. In 1901 Royal presented the Missouri law in Indiana. The Indiana legislature did pass a baking powder bill during that session, but it did not prohibit the use of alum. It prohibited cream of tartar in baking powder. But Royal's supporters were not going to allow this bill to become a law. In an act of guerrilla warfare, before the bill was signed by the governor, the words "bitartrate of potassium" (cream of tartar) were scratched off. So what the governor signed into law was meaningless. Colonel William W. Huffman, principal secretary of the Indiana senate, was arrested in the statehouse and charged with mutilating a law; the penalty was prison, a fine, or both. He received neither, because the judge ruled that the bill, although passed by the legislature and signed by the governor, was not a law until it was enrolled by the secretary of state.[16] Since it had not been enrolled, it was not a law. Huffman had erased the words and destroyed the bill, but he had not mutilated a law. He was acquitted.[17] Royal did not introduce anti-alum legislation in Indiana again.

Perhaps Royal suspected that the Hulmans had had something to do with the failed Indiana law prohibiting cream of tartar in baking powder, because soon after, Royal repeatedly targeted Clabber and Hulman & Co. specifically. One Royal

Table 7-1. Contested in the Baking Powder Wars

	1900	1901	1902	1903	1904	1905
Arkansas		✓		✓		✓
California				✓	✓	✓
Colorado				✓		
Delaware		✓				✓
Florida						✓
Georgia	✓	✓				✓
Idaho	✓			✓		✓
Illinois	✓	✓				
Indiana		✓				
Iowa	✓				✓	
Kansas		✓				✓
Kentucky	✓				✓	
Louisiana			✓			
Maine						✓
Maryland	✓					
Massachusetts	✓	✓	✓	✓	✓	✓
Michigan		✓				✓
Minnesota	✓	✓	✓	✓		
Mississippi	✓					
Missouri	✓	✓	✓	✓	✓	✓
Montana		✓				✓
Nebraska		✓		✓		
New Hampshire		✓				
New Jersey	✓	✓	✓			✓
New Mexico Territory		✓		✓		
New York	✓	✓	✓	✓	✓	✓
North Carolina	✓					✓
Ohio	✓					
Oregon				✓		✓
Pennsylvania	✓			✓		✓
South Carolina	✓					
South Dakota		✓				✓
Tennessee		✓	✓	✓	✓	✓
Texas		✓				✓
Utah		✓		✓		✓
Vermont						✓
Virginia	✓					
Washington		✓		✓		✓
West Virginia		✓		✓		✓
Wisconsin		✓		✓		✓
Wyoming					✓	✓
Washington, D.C.	✓	✓	✓	✓	✓	
TOTALS	18	24	8	18	9	26

Source: Compiled from Minutes of ABPA meetings, 1900–1905, BPC, 965–1024.

article, titled "Alum Baking Powders—Congress Acting to Suppress Their Sale," said that the state senate Committee on Manufactures would recommend that alum baking powders be prohibited. First on the list: Clabber, "Manufactured by Hulman & Co., Terre Haute."[18]

Hulman and the other baking powder companies were powerless to defend themselves against articles like this. In January and February 1901 alone, editors from seven newspapers in Indiana, Illinois, and Tennessee wrote to Hulman & Co. saying they could not print articles that Hulman had submitted rebutting Royal's accusations, because their contracts with Royal prohibited it. One editor said he would print the Hulman ad in spite of his Royal contract. Another editor pinpointed how much power newspaper advertising had: "There is no question about it, the prejudice aroused by newspaper advertising is the only thing which holds up the cream of tartar powders."[19]

The problem was that revenues from local advertising were usually not enough to support small-town newspapers. Advertising dollars from large national companies like Royal kept many local newspapers throughout the United States afloat. Even if local journalists knew about corruption and wanted to write about it, companies that bought legislators also bought newspaper advertising. Business managers overrode editors.

Even though Royal was successful in controlling newspapers, it was less successful in the courts. In 1900, *Grocery World*, a trade publication favorable to the American Baking Powder Association, reported a failed attempt to prosecute a grocer in Minnesota for selling alum baking powders.[20] The Minnesota Supreme Court dismissed the case and suggested there had been collusion between Royal and pure food officials. In an extraordinary intervention, the governor of Minnesota injected himself into the debate. He "instructed the Pure Food Department to take no part in the controversy whatever." He also said, "If baking powder is labeled according to the law, the consumer should be allowed to make his own choice, without the meddling of the Royal Baking Powder Trust."[21]

The American Baking Powder Association, too, relied on this caveat emptor argument—let the buyer beware. They countered Royal's attempts to make alum baking powders illegal by saying that as long as the baking powder was labeled, it was the buyer's decision. Caveat emptor applied to drugs as well as to food, because there were no illegal substances in the United States at that time. A hearing on baking powder in Massachusetts drew this analogy: "The use of morphine has steadily increased, and many people carry with them the pretty little hypodermic outfits, with which an injection can be so easily made."[22] If people wanted to shoot morphine, that was their decision. Similarly, they should be allowed to choose their own baking powder. It was a laissez-faire opinion: the state had no place in making the decision or prohibiting the product; its only role was to make sure the consumer was not defrauded.

Meanwhile, in Missouri, one of the businessmen who had been arrested for selling baking powder, Whitney Layton, took his case to trial. Layton's attorneys drew a parallel between baking powder and "the oleomargarine cases." It was clearly understood by that time that the purpose of the law against oleomargarine "was not to protect the public health, morals, or welfare, but to foster one industry at the expense of another lawful one, which was prohibited."[23]

A wide range of witnesses, from boardinghouse owners and hotel managers to chemists and physiologists, testified on Layton's behalf that alum baking powder was safe and effective. Physicians swore that they "had never known of any case where the digestion or health of a person had been injuriously affected by food made with an 'alum' baking powder."[24] Even the scientific experts for the prosecution stated under cross-examination that they had "no knowledge of any recorded instances in which functional disorders or disease or impairment of the digestion and general health had resulted to any human being from the use of alum baking powder as an ingredient in the preparation of food."[25]

In the trials against the manufacturers of alum baking powders and the grocers who had sold it, the attorneys who prosecuted the case were not state of Missouri prosecutors; they were employed by Royal. Likewise, many of the witnesses who testified on behalf of the state were paid by Royal, fifty to one hundred dollars per day plus expenses, an enormous sum at the time.[26]

The judge ruled that although "I am unable to find in the evidence in this case any just ground for a ruling that alum baking powders, of themselves, when used in the preparation of food are in any wise less wholesome than any other variety of baking powders," still, the defendant had broken the law. Layton was found guilty and fined one hundred dollars, the minimum allowed under the law.[27]

Layton appealed to the Missouri Supreme Court. The state argued that it was within its constitutional powers to declare substances illegal. It cited the case of *Mugler v. Kansas*, in which a brewer continued to make beer after Kansas passed a law making it a dry state. The Missouri Supreme Court upheld the lower court's guilty verdict against Layton.[28]

The case then went to the U.S. Supreme Court on Fourteenth Amendment grounds. Layton claimed that he had been deprived "of his liberty and his property without due process of law." On December 22, 1902, the Supreme Court dismissed the case because the constitutional question involved "was not raised in and submitted to the trial court."[29]

What was the impact of the baking powder law on the people of Missouri? Total sales of all types of baking powder in Missouri at that time were five million pounds per year. Royal's sales accounted for only five hundred thousand of those pounds.[30] An 1889 report by social workers for the State Bureau of Mines and Mine Inspection reveals how deeply baking powder had penetrated even in remote areas. The social workers went to coal miners' homes and documented their living conditions, including the food and food budget. Miners made $1.50 per day. They were paid in scrip and forced to shop at the company "pluck me" store at artificially inflated prices.[31]

Baking powder is the only ingredient for which the social workers recorded brand names. Royal was aggressive in entering into contracts with the mining companies, because that is the brand name that appears most often. Baking powder ranged in price from twenty cents per pound to sixty-five cents per pound. Royal was usually at fifty cents. Dr. Price's baking powder was forty-five cents. These prices were double what either alum or cream of tartar baking powders sold for in stores where there was competition. All baking powders, even the cheapest, cost more than baking soda, which was five to ten cents per pound; beans, three cents per pound; bacon, ten cents; and coffee, twenty-five to thirty cents.[32]

The social workers also included recipes in their reports. There are recipes for biscuits, corn bread, and cookies, all leavened with either baking powder or baking soda. Housewives and women who operated boardinghouses already knew how to make the other foods they cooked, so they did not need recipes for them. But foods with chemical leaveners were new. It also indicates that these women were literate.

In 1900 the ABPA had to shift its energies from Missouri to Washington, DC. On March 15, 1900, Senator Redfield Proctor of Vermont (R), chairman of the Committee on Agriculture and Forestry, introduced Senate Bill 3618. Section 7 of the bill gave ultimate authority, with no recourse and no appeal, to the secretary of agriculture to "fix standards of food (including baking powders) and to determine the wholesomeness of substances added to foods." In practical terms, this meant that Harvey Washington Wiley, MD, the USDA chemist since 1883, would decide food standards.

Wiley had come to the USDA after studying chemistry in Germany, obtaining a medical degree in the United States, then a BS in chemistry from Harvard after six months of intensive study and nine years as professor of chemistry at Purdue University in Indiana. Wiley's predecessor had proposed a pure food law in 1880; in the next twenty-five years, more than one hundred bills were introduced. Wiley had backed them all.[33] He was devoutly anti-alum. The ABPA immediately understood: "What can this bill mean but that the alum baking powder industries are to be destroyed?"[34]

Over several days in April 1900, members of the ABPA testified at length before the committee. They made three main points: that alum baking powder was (1) wholesome, (2) economical, and (3) made from American products. Cream of tartar, on the other hand, was composed of 85 percent foreign ingredients. The ABPA portrayed Royal as old world and old-fashioned, unable "to keep abreast of scientific progress," and opposed to "the most recent advances in science."[35] In other words, it was un-American.

In their testimony, all members of the ABPA repudiated the idea that they were in a trust, or that they had been formed as a business combination to regulate prices or trade in violation of the 1890 Sherman Anti-Trust Act: "We are . . . simply an association organized for defensive purposes."[36] They had prepared carefully and stated their position clearly: they wanted pure food legislation as long as it was fair, meaning it did not exclude alum baking powders. They spoke about the adverse impact of Royal's advertising, Royal's intention to become a monopoly, and the Missouri law. They presented scientific testimony that alum baking powders were wholesome.

The president of the General Chemical Company and the presidents of the Detroit Chemical Works and of the Provident Chemical Works of St. Louis also testified on behalf of alum baking powder. Some of the men said they had been eating baking powder biscuits for decades and were perfectly healthy. They were all careful to speak in factual, respectful terms to the committee. Only R. B. Davis, of Davis Baking Powder, headquartered in Hoboken, New Jersey, let slip his animosity to the bill and to Wiley personally: "These gentlemen in the Department of Agriculture . . . know as little about baking powder, practically, as a cow knows about algebra."[37]

In Missouri in 1901, at the next biennial session of the legislature, consumers put pressure on legislators to repeal the alum bill. However, William Ziegler and Daniel J. Kelley got on the train in New York City and headed for Missouri. Kelley was principal, president, and publisher of the *American Queen*, a monthly grocery trade paper. He lived at the Plaza Hotel. He was also William Ziegler's right-hand man.[38] Ziegler told acquaintances that he was going to Missouri "to crush out the alum baking powder people."[39] And so he did.

The Missouri House of Representatives voted 109 to 5 to repeal the anti-alum law. The state senate, bribed again, killed the repeal. But the enraged Missouri house was on the war path. It passed a resolution condemning the Missouri senate, because "for the next two years every family in the State will be compelled to buy high-price [*sic*] Trust baking powder, and every merchant handling other than Trust goods will be liable to criminal prosecution."[40]

The Missouri state senators were easy to bribe, but they weren't cheap. Many senators were already on the payrolls of other businesses for five hundred dollars per month, the equivalent of a male factory worker's annual salary. Kelley later said, "It was the highest priced legislature I ever ran up against."[41] On March 19, 1901, the day after the Missouri General Assembly ended, Kelley withdrew seven thousand dollars from the American Exchange Bank in St. Louis and gave it to Lieutenant Governor John A. Lee. Lee then went to the Laclede Hotel and gave the money to Senator Frank Farris, who put one thousand dollars in each of seven envelopes.[42] The money was referred to as "boodle." It was also called, in an early use of the term, a "slush fund."[43]

Consumers continued to pressure the Missouri legislature. By the time the legislature convened again, in 1903, more than five thousand merchants had signed a pledge stating that they would vote only for candidates who promised to repeal the anti-alum law.[44] Kelley was in Missouri again, registered at a hotel as "Mr. Brown." Repeal failed again, but in a way that aroused suspicions. Lieutenant Governor Lee tried to play both sides in the baking powder war. When he realized how strongly his constituents felt about alum baking powder, he met with alum baking powder representatives, apologized for opposing them, and vowed to help them. All the while, he was taking Royal's money.

Lee's overtures to the alum baking powder contingent angered Farris. Farris got his revenge in 1903 when repeal of the alum law came up for a vote. He deliberately left the state senate chamber and took two other senators with him. This left the remaining senators tied, 16–16. Farris's move forced Lee into the spotlight. As president of the senate, Lee had to cast the deciding vote. He wanted to preserve his reputation as a crusader for reform and use it to propel him into the governorship; he also knew that his constituents wanted the law repealed. But he had taken Royal's money. While Lee wavered and waffled, "Farris seemed to take particular delight in the situation."[45] Lee's greed finally prevailed over conscience, and he cast the deciding vote against repeal.[46] Ironically, Farris later said that how Lee voted was inconsequential. If Lee had voted for repeal, Farris would have called the other two senators back and demanded a new vote. If that had happened, Lee might have kept his reputation as a reformer intact and become the next governor of Missouri. The senators flaunted the bribe money by asking for change for thousand-dollar bills in saloons.[47] Rumors spread.

The baking powder trust scandal broke in stages. That it broke at all was because between 1899 and 1903 profound changes occurred in American media. America's transportation and communications infrastructure spread the Missouri bribery from local curiosity to national scandal. Industrialization had reached entertainment, and Americans at the turn of the twentieth century, with a median age of 22.9, were eager for the new.[48] Millions traveled to the Columbian Exposition in Chicago in 1893 and the Louisiana Purchase Exhibition in St. Louis in 1904 to see new technology, people, and foods from around the world. The railroad took them to the excitement at the fair; when they returned home, it delivered the excitement of the new journalism to their homes. In the cities a new type of media was growing. A nickel bought a few seconds of watching pictures that moved: a man sneezing, a couple kissing, firemen rushing. Then, as now, print media had to be stimulating to compete with the visual.

The Gilded Age saw an explosive growth of newspapers and magazines. From a few hundred in 1860, in 1900 there were 2,190 daily newspapers; 15,813 weekly newspapers; and 5,500 periodicals in the United States.[49] Businesses advertised heavily in them: "By 1891 magazines with one hundred pages of advertising were commonplace."[50] The old scholarly literary magazines—*Harper's*, the *Atlantic*, *Scribner's*, and *Century*—were displaced by new popular magazines with exciting stories about modern urban life. Circulation for the older magazines had never exceeded 130,000 per year; the new magazines had annual circulations from 350,000 to more than a million each.[51]

Ten-cent muckraking magazines like William Randolph Hearst's *Cosmopolitan* were the upscale urban version of the dime novel, with reality that exceeded fiction—the early twentieth-century equivalent of reality TV. The adventures of fictitious characters like detective Nick Carter, inventor Frank Reade Jr., and outlaw Deadwood Dick were joined by thrilling true stories: George Kibbe Turner's study of prostitution, gambling, and liquor in Chicago; how New York City gangster "Kid" Twist provided repeat voters for corrupt Tammany elections; the travails of working women, written by women journalists such as Bessie and Marie Van Horst; novelist Frank Norris on speculation in wheat; real policemen talking about lawlessness in the United States; and more in *McClure's*, *Leslie's*, *Everybody's*, *Munsey's*, *Collier's*, and other new magazines.

Unlike the anonymous writers who cranked out dime novels in sweatshop piecework conditions, the muckrakers were a new breed: celebrity journalists. They had bylines and were well paid. Lincoln Steffens commanded two thousand dollars per article and had "standing invitations to talk in nearly every town in this country."[52] He was so famous, a cigar was named after him.[53] Upton Sinclair, author of *The Jungle*, the stomach-turning novel about the Chicago meatpacking industry, said that these new journalists reported on the things "everybody knows" and that Americans saw going on around them: the corruption, pollution, and adulteration that had become standard business practices.[54]

A native of Sacramento, California, Steffens was one of the only muckraking journalists who had not attended Knox College in Galesburg, Illinois. The Presbyterian college was founded in 1837 as an anti-slavery institution. Imbued with evangelical Christian fervor, these Midwestern muckrakers, transplanted to New York City, believed that one of the great evils was the "unjust accumulation of wealth."[55] William Allen White, whose mother was a Knox alumna, equated "the bondholding aristocracy" with "the slaveholding oligarchy."[56] The muckrakers were looking for "a John Brown, who arouses public sentiment against these [political and social] evils."[57] Lincoln Steffens obliged; baking powder was his platform in the moral war.

The first article about the baking powder scandal appeared on December 2, 1902, because a persistent reporter for the *St. Louis Post-Dispatch*, J. J. McAuliffe, found a willing ear in Joseph "Holy Joe" Folk, the Progressive crusading circuit attorney who had prosecuted the boodlers for corruption in the St. Louis transit system.[58] Folk had gained national attention because Steffens had praised Folk in his "Shame of the Cities" series in *McClure's* magazine.

Sam McClure, Knox alumnus and owner of *McClure's*, saw himself as a king maker, with his sights set on the presidential election of 1908, after Theodore Roosevelt finished his second term. The Irish-born and Indiana-raised McClure went back to his Midwestern roots for the king he wanted to make: "Holy Joe." McClure instructed Steffens to go back to Missouri and escalate his series on corruption in the cities to corruption at the state level, to find something that would propel Folk into the governor's mansion and then make him a presidential candidate.[59]

Before Steffens could get back to Missouri, the *St. Louis Post-Dispatch* revealed that the prestigious and official-sounding National Health Society, which had repeatedly condemned alum baking powders, and just as repeatedly denied that it was connected to Royal, had a total of three members. Its president and treasurer was Daniel Kelley, Ziegler's right-hand man; its secretary was Joseph J. Cunningham, Royal's attorney.[60] The third member was United States senator and former Missouri governor William J. Stone. On April 20, 1903, in Kansas City, Stone used the "best defense is a good offense strategy." He addressed the Missouri Democratic Press Association and fulminated: "I hope God will wither my hand, palsy my tongue, and burn my heart in the flames of hell before I will intentionally dishonor any position to which the people of Missouri assign me."[61] His speech was effective. Stone went on to serve in the U.S. Senate until his death in 1918.[62]

The baking powder scandal widened in the spring of 1903 when the *St. Louis Post-Dispatch* printed a series of articles. On April 7, at the request of Missouri authorities, Daniel Kelley was arrested in New York. However, he was released on his own recognizance. Lee sent Kelley a thinly disguised cloak-and-dagger telegram: "Your health being poor, brief . . . trip, if taken, would be greatly beneficial." Lee signed it with an alias, "James Sargent." Kelley got the message and had a sudden urge to see springtime in Quebec.[63]

It was because of Holy Joe Folk that Kelley then, like fugitives now, had to go to Canada. Folk had gotten tired of bribed officials taking their ill-gotten American dollars to Mexico and living like royalty. He appealed to Secretary of State John Hay, famous for negotiating the Open Door policy that allowed the United States access to China markets, to amend the United States' treaty with Mexico to include bribery in the list of extraditable offenses. Hay said it could not be done. Folk went over Hay's head to President Theodore Roosevelt, who thought it was a bully idea and made it happen. John Hay had opened the door to China, but Holy Joe Folk closed the door to Mexico.

However, Kelley, an experienced bribe giver, knew enough not to trust a bribe taker. Plus, he despised Lee. He wrote a threatening letter to Lee: "I have in my possession every letter you ever wrote to me and every paper you have put your name to. If you have or should malign me . . . these documents will be used and will protect me fully."[64] When Lee did try to save himself by throwing Kelley to the press wolves, Kelley made good on his promise and showed sixty letters to the persistent McAuliffe, the *St. Louis Post-Dispatch* reporter who had tracked Kelley down in Canada.[65]

The letters were very damaging. They revealed every step of how Royal had bribed the Missouri senate and identified every palm that had been crossed. On April 9, 1902, Lee had written to Kelley that the price for stopping legislation in the Missouri House of Representatives to repeal the alum law would be fifteen thousand dollars. On another occasion, Lee asked for twenty-five thousand dollars for his gubernatorial campaign. On January 3, 1903, Lee wrote to Kelley that he had gotten into a dispute and wanted to "shut his [opponent's] mouth with some money."[66]

Lee perceived that his problem was not that he had engaged in bribery and betrayed his constituents and his oath of office, but that the press had found out and would not stop writing about it. Lee wrote to Kelley and asked for help in silencing the *St. Louis Post-Dispatch*: "Why not give the paper a chunk of business? Maybe that will stop it. Unless you do it, I am out of the race for governor."[67]

Lee overvalued his bargaining position and underplayed the difficulty he was in. After the scandal broke, Lee made a delusional statement to McAuliffe: "I will try to live it down. I believe I can." Shortly afterward, Lee sent his wife to negotiate with Missouri attorney general Edward C. Crow to see if Crow would confine the investigation to baking powder and not inquire about corruption in general in the legislature. At the same time, Lee was negotiating with Colonel Phelps, the railroad lobbyist, for a deal that would keep him from testifying at all. To protect their own positions in the Missouri government, Phelps and another lobbyist were willing to pay Lee a thousand dollars a month for two years to stay away and not testify. By then the statute of limitations would have run and they would all be immune from prosecution. But greedy Lee ridiculed the offer and demanded one hundred thousand dollars.[68] There were no further negotiations.

Finally, in front of the grand jury on April 15, Lee realized how much trouble he was in. After ten minutes he collapsed, blubbering.[69] Then he took the advice he had given Kelley and disappeared. On April 18, 1903, the *Los Angeles Herald* ran an article on page one, above the fold:

LEE HAS VANISHED
Lieutenant Governor of Missouri Wanted
BOODLING INVESTIGATION IS
DELAYED BY HIS ABSENCE[70]

Lee returned and testified before the grand jury in St. Louis on April 24, 1903. Three days later he resigned his office, "the first high official in Missouri who ever has quit under fire."[71] Under indictment, he cried in open court and named names. Two senators were jailed for contempt from April 27 to 29 for refusing to tell the grand jury where they got the thousand-dollar bills. On May 12, Lyons, the senator who had introduced the anti-alum bill, took the stand and the Fifth Amendment.[72]

Then Missouri went after Ziegler and Kelley. In July 1903 the State of Missouri indicted William Ziegler for "feloniously and corruptly devising, contriving, scheming and intending to induce and procure . . . the said Frank H. Farris, to prostitute, abuse and betray his trust and violate his duty as a member of the said State Senate . . . of the State of Missouri, towards the people of the State" to prevent the passage of the repeal bill.[73] The state claimed that Ziegler was a fugitive from justice.

On September 17, Missouri governor Alexander Monroe Dockery put a five-hundred-dollar bounty on Daniel Kelley's head.[74] Kelley knew that if the *St. Louis Post-Dispatch* reporter McAuliffe could find him in Canada, so could the Pinkerton bounty hunters.[75] In February, Kelley got a sudden urge to see winter in London and set sail.[76]

On November 16, 1903, the governor of Missouri contacted the governor of New York and formally requested that he turn Ziegler over to Missouri authorities. Ziegler's attorneys had seen this coming; they demanded an extradition hearing. Missouri sent its attorney general, Edward C. Crow, and Judge Thomas B. Harvey to Albany to argue its case at Ziegler's extradition hearing on December 7, 1903. The Missouri brief said that they were "in deadly earnest," because bribe givers were as "wicked" as murderers, and "The man who bribes the official aims at the assassination of the commonwealth itself."[77] At one point Ziegler was so upset at the charges Crow leveled at him that he "retired to a corner of the room and wept bitterly." Ziegler claimed the charges were based solely on malice, and "his whole body shook with emotion as he sat sobbing."[78]

In his defense, Ziegler produced alibi witnesses, primarily relatives and employees, who contradicted every public admission and boast that Ziegler had made about being in Missouri when the bribery took place. Every one of them swore that Ziegler had been in New York. On February 1, 1904, New York governor Benjamin Odell ruled that Ziegler was in New York City when the bribery occurred and therefore was not a fugitive.[79] Odell refused to extradite Ziegler, a rare response to an extradition request from another governor.[80]

A national bomb burst in the baking powder war in April 1904. Lincoln Steffens's article "Enemies of the Republic," about Royal's bribery of the Missouri state legislature, appeared in *McClure's*. Steffens attacked the Royal Baking Powder Trust and the lobby as an "extra-constitutional government."[81] He quoted Folk as saying, "Bribery is treason, and a boodler is a traitor."[82] Steffens interviewed Ziegler for the article and found him "a generous, able, courageous, and rather humble-minded man." He also said that Ziegler had not the least understanding that his bribery of the legislature in Missouri had changed the government by creating a "third house"—the lobby—that was not a representative government.[83] Steffens subtitled his article "Business as Treason."

A month after the publication of the Steffens article there was another showdown, this time between Royal and the American Baking Powder Association. On May 18, 1904, the U.S. Government Committee on Food Standards held a hearing on baking powder standards in New York City. Royal and the ABPA both wanted the coming federal law to be based on the strengths of its own product. Royal wanted the standard to be "wholesomeness." The APBA wanted a standard based on how much leavening power a baking powder had. Each standard would exclude its competitor's product.

Royal sent two representatives to the hearing, Professor William McMurtrie and Henry La Fetra. McMurtrie, Royal's head chemist, had served as chief chemist at the U.S. Department of Agriculture (USDA) and then became professor of chemistry at the University of Illinois before he joined Royal. Accompanying McMurtrie was the head of Royal's advertising department, Henry La Fetra, who had been with Royal since 1882.[84] La Fetra had a grandiose personality; since 1899 he had lived in a seventy-three-hundred-square-foot mansion in New Jersey, one of only two private residences designed by Henry Bacon, the architect of the Lincoln Memorial.[85]

The association's strategy at the hearing was two-pronged. They presented facts about baking powder, but like good poker players, they also played their opponents: "Knowing Mr. La Fetra to be extremely peppery, we felt that if we could get him excited, we would have him at a disadvantage." The ABPA baited La Fetra, and he bit hard—repeatedly. To irritate him initially, the ABPA pretended a show of generosity and urged the committee to set the standards for baking powder leavening ability low enough to keep Royal in the market. The ABPA, sarcastically feigning solicitude, did not want Royal, a "two-spoon powder," to have to label itself "half strength" or "inferior." That got La Fetra outraged and out of his chair. He declared that Royal was "not in need of any charity." The ABPA continued its condescension, stating that they "did not wish to have a standard fixed so high as to be embarrassing to our friends."[86]

La Fetra exploded repeatedly during the hearing. At one point he said that the U.S. Government Committee on Food Standards, whose job it was to fix food standards, should not fix any standards at all. Morrison reported to the ABPA, "We suppose McMurtrie wanted to kick La Fetra." Then La Fetra pointedly told Harvey Washington Wiley, chief chemist for the USDA, that "the experience of the Royal Baking Powder Company with 'Chemists—Official Chemists'—had not been satisfactory." Morrison again observed, "We were of the opinion that McMurtrie wanted to kick La Fetra again after this." The end result of the meeting was that the ABPA had behaved respectfully and was on the record as being in favor of standards, while Royal had made a poor showing.[87]

During this time, attorneys for the defendants in the Missouri bribery case were doing anything they could to delay the trials. They repeatedly requested continuances because of illnesses of defendants and witnesses, and because defense attorneys needed more time for preparation. They also tried to get changes of venue.

In the meantime, the press all over the United States printed many sharp cartoons about the bribery. They showed the boodlers as jailbirds and pickpockets. There was even a cartoon of Colonel Phelps lounging behind the curtain in the Missouri senate chamber, pulling the strings of his senator puppets.

Royal tried unsuccessfully to repeat its bribery of the Missouri legislature in other states. Legislators were arrested in Arkansas and California for taking bribes to ensure passage of an anti-alum law. Their arrests ensured that the law did not pass.

Royal's attempts to use lawsuits in other states to stop the sale of alum baking powders also failed. In November 1903 the Los Angeles Board of Health ordered that anyone selling alum baking powder should be arrested. Two cases went to trial. The difficulties in prosecuting new food technology became apparent in this

The Lobby pulls the strings in the Missouri Senate. *St. Louis Globe-Democrat*, April 13, 1903.

The Missouri baking powder scandal blows up. *New York Herald* [n.d.].

lawsuit. The Los Angeles court struggled to define exactly what baking powder was. The court looked first to the dictionary for the definition of "baking powder," then for the definition of "food," and then asked, "Is baking powder a substance which is fed upon to support life by being received within and assimilated by the organism of man? Is it eaten for nourishment? Does it supply nourishment to the animal organism? It is nutriment or aliment?"[88] Attorneys for the defense argued that alum was not food. The court agreed and decided that baking powder was not added to food to provide nutrition and that it dissipated in cooking; ergo, it was not food.[89] And since it was not food, it could not be adulterated. The defense attorneys also made it clear that business competition was behind the prosecution, with the intent to force consumers "to purchase the product of the Royal Baking Powder Company, the Baking Powder Trust, at a price of forty-five to fifty cents a pound, instead of sixteen cents a pound, which is the price of the so-called alum powders."[90]

In September 1904, a few months after the hearing with La Fetra, the members of the American Baking Powder Association met in St. Louis to attend the

Thousand-dollar bribery bills are the dough. *St. Louis Globe-Democrat,* April 10, 1903.

Louisiana Purchase Exhibition world's fair and enjoy the fruits of the Steffens article. For the first time, the ABPA was optimistic. They believed that because Royal had been exposed in the national press, it could not use the same tactics again. In an address before the International Pure Food Congress in St. Louis that month, Morrison called the baking powder controversy "one of the most remarkable commercial contests which has taken place during the last five years or at any other period. I think it is only comparable perhaps to the Standard Oil contest. It differs from that struggle in that the smaller manufacturers have in our case won the fight."[91]

The ABPA had won. On January 6, 1905, the first item up for discussion was Missouri House Bill Number 1, repeal of the anti-alum law. On February 9 the house voted for repeal, 123–1. A repeal bill was also introduced in the senate, where it passed 28–4. Senator Frank Farris led the dissenters.[92] On February 21, 1905, Missouri's new governor, Joseph Folk, signed the bill into law.[93] After six years, the Missouri baking powder law had finally been repealed. By then the cream of tartar baking powders, which had begun fifty years earlier with almost the entire

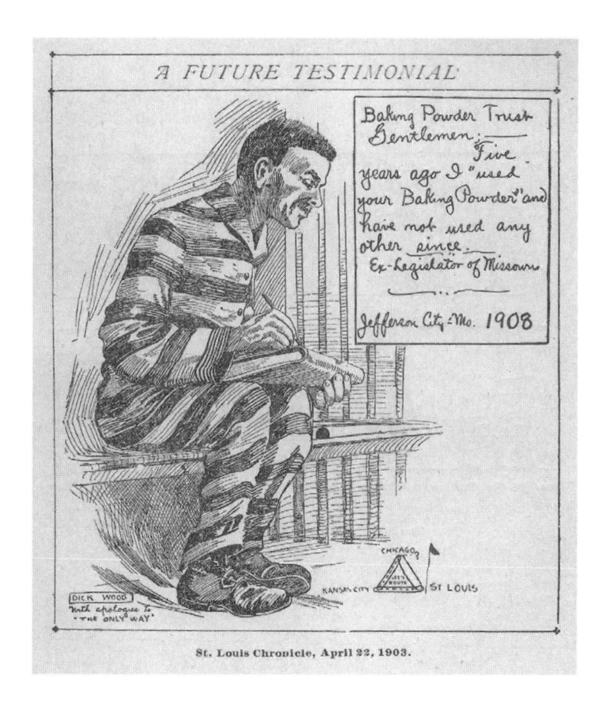

This cartoon, which seemed to be such a certainty in 1903, proved to be wishful thinking by 1905. *St. Louis Chronicle,* April 22, 1903.

baking powder market and which had controlled one-fifth of the market in 1899, had dropped to one-sixth.[94]

By August 1905 it was all over in Missouri. On August 6 Frank Farris was acquitted by a jury split 8–4.[95] Farris was the linchpin: if he was not guilty of handing out bribes, then the other senators could not be guilty of taking bribes; indictments against them were dismissed.[96] So was the indictment against Lieutenant Governor Lee.

What of Colonel Phelps, the lobbyist sitting behind the curtain and pulling the Missouri Senate's strings on behalf of the Missouri Pacific Railroad? After the grand jury indictments, Phelps "retired to parts unknown" and was universally acknowledged to be politically dead.[97] Five years later, he was elected to the Missouri House of Representatives, where he served two back-to-back terms from 1910 to 1914, when he was elected to the Missouri Senate. After thirty years of lobbying for the railroads, Phelps was reincarnated as "a relentless foe of the railroads, forcing through the legislature some of the most radical anti-railroad legislation ever enacted in any state."[98] But maybe Colonel Phelps left an even more important legacy. Perhaps a young newspaperman in the Midwest, L. Frank Baum, had Phelps in mind when he wrote about that other man behind the curtain, *The Wizard of Oz.*

Also in 1905, *Cosmopolitan* magazine serialized *The Jungle*. Upton Sinclair's exposé of the Chicago meatpacking industry gave a sense of urgency to the national Pure Food and Drug Law that Congress had been debating in earnest for the preceding six years. On June 30, 1906, Congress passed the Pure Food and Drug Act. The part of the law that pertained to baking powder stated: "An article shall be deemed to be adulterated . . . if it contain[s] any added poisonous or other added deleterious ingredient which may render such article injurious to health."[99] Since both cream of tartar and alum had scientists who could testify that their products did not contain *added* poisonous or deleterious ingredients and was not injurious to health, both were permissible. Because both sides also had scientists who could testify that their competitors' products were poisonous or contained deleterious ingredients, the Pure Food Law became more ammunition in the baking powder war.

William Ziegler did not live to see his hopes of a national law excluding alum baking powders dashed by Congress. On May 25, 1905, he died from complications of injuries he had suffered the previous October in a carriage accident near his Connecticut home. The ABPA minutes included a notice about Ziegler's death along with a prophetic remark he had made: "It is said that he loved power more than money; it is said that he declared that he would suppress our industry, even if it cost him his life or his fortune."[100]

In addition to baking powder, much of Ziegler's fortune came from real estate. Beginning in 1886, he had started buying lots by the thousands in New York: two thousand at Morris Park; fifteen hundred in Flatbush and New Utrecht; twenty-five hundred in Flushing and Corona; two thousand on Staten Island; and "much other property on Liberty and Cedar streets and Fifth and Madison Avenues." He also owned six thousand lots in Linden, New Jersey.[101] In 1902 he had purchased Great Island in Darien, the Connecticut estate where he suffered his accident. More than two hundred laborers made extensive improvements on the 170 acres, creating a polo field, a yacht basin, numerous outbuildings, and a beach in addition to the mansion itself.[102]

Ziegler's death and the fortune he left behind precipitated lawsuits that set legal precedents and aired much dirty family laundry. He left $16.5 million to his son, William Jr. The son was really Ziegler's nephew, Conrad Brandt. Like Julius Caesar and his nephew Augustus, Ziegler had adopted the boy and renamed him so that he would have an heir for his empire. Ziegler had also adopted Conrad's sister, Florence Louisa Brandt, but later reversed the adoption. When Ziegler died, Florence sued to have the reversal of the adoption nullified on the grounds that her natural mother had not consented to it. This was the first time in American law that an adopted person who had been legally un-adopted had attempted to get re-adopted. Florence further claimed that Ziegler had not wanted to reverse the adoption but that his wife, Matilda, had "early formed a feeling of ill-will toward me, which gradually grew to intense hatred," and had forced him to reverse the adoption. Florence lost her suit in part because of the institutionalized misogyny at the time. The court ruled that the consent of Florence's mother had been dispensed with on the grounds that she was an adulteress and therefore had been "deprived of parental control."[103]

Matilda brought her own lawsuit against her husband's estate. In contrast to the millions Ziegler left to his son, he bequeathed only fifty thousand dollars a year to his wife. Matilda contested the will on the grounds that Ziegler was insane. This was news to his friends, although some of them admitted that Ziegler's obsession with repeatedly equipping expensive expeditions that failed to find the North Pole might have been a sign that he was becoming feebleminded. In the end, Matilda Ziegler received $1.2 million in cash and $1.3 million in Royal stock.[104]

William Ziegler's death was not, as the ABPA had hoped, the end of the baking powder war. Royal and William Ziegler Jr. continued Ziegler's campaign against alum baking powders. In 1899 Calumet's William Wright had elevated his twenty-four-year-old son, Warren, who had been working at the company for nine years, to secretary of the company. Like an American version of the Charles Dickens novel *Bleak House*, about an all-consuming, multigenerational lawsuit, both sons zealously continued into the twentieth century the baking powder war their fathers had begun in the nineteenth.

CHAPTER 8

The Alum War and World War I

"What a Fumin' about Egg Albumen," 1907–1920

The wonderful progress of the United States, as well as the character of the people, are the results of natural selection.
—Charles Darwin, 1871

The requirements upon modern business life decree that only the fittest shall survive.
—Calumet Baking Powder, *Why White of Eggs*, 1914

"Survival of the fittest." By the second decade of the twentieth century, the complex scientific Darwinism of the mid-nineteenth century had been reduced to a catchphrase of Social Darwinism. For baking powder companies this meant that after the Pure Food Law was passed in 1906, the baking powder war not only continued but expanded as well. The war was fought increasingly on the national level, in the new national press and in new federal agencies. In the western United States, the alum companies turned on one another in a war that echoed the cream of tartar war of the 1880s and 1890s.

Royal and Calumet continued to wage the baking powder war through advertising and direct attacks on each other. They also brought it to a new medium: the baking powder can. By 1909, printed on the Calumet can, at the top above the name, was "Not Made By The Trust." In 1914 Royal countered with a slogan in its advertising: "Royal contains no alum—Leaves no bitter taste."[1]

In addition to attacking each other, the labels of both companies also laid claims to being "natural." The Calumet can was emerging out of an eggshell like Botticelli's Venus rising from the sea, a visual reference not only to nature but also to baking powder's ability to replace eggs as leaveners. The can is surrounded on both sides by golden wheat, furthering the natural theme. Royal's image embodied its

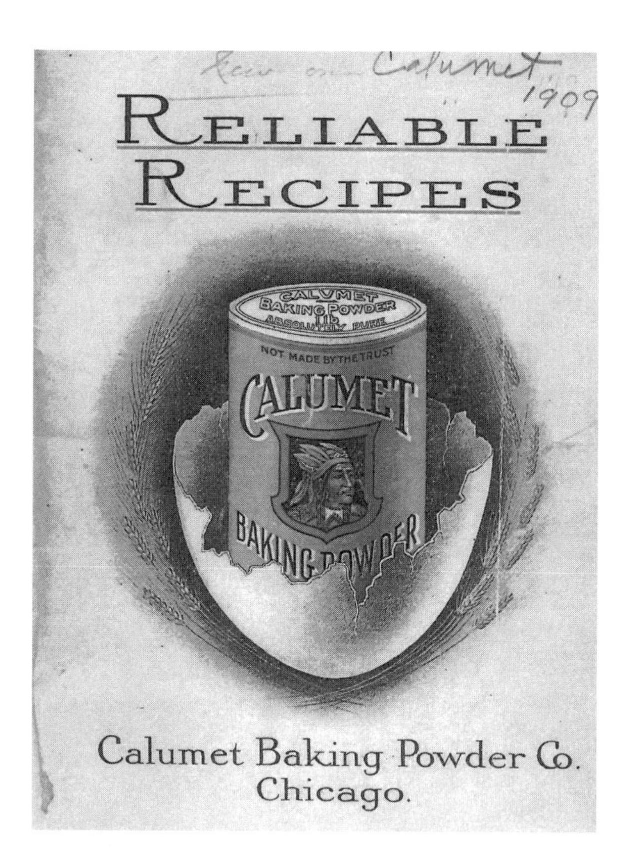

"Not Made By The Trust" was later replaced with "Not Made by a Trust." Kraft Heinz Company.

claim about originating with grapes. In other ways, the cans of both companies complied with new labeling laws.

At the beginning of the twentieth century, all baking powder companies had to change how ingredients were listed on their labels to comply with new labeling laws in Wisconsin, Minnesota, and other states. In its July 10, 1905 report, the Minnesota State Dairy and Food Commission issued specific guidelines for labeling baking powder. The intent of the law was to cut through the chemical confusion and make the labels understandable to the consumer:

> The label stating the ingredients used in baking powder must be printed on a separate and distinct white or light colored label, and must be immediately followed by the manufacturer's name and address. All advertising matter must be eliminated from the label, and this department rules that the names of the ingredients must be absolutely intelligible.
>
> Technical terms for ingredients will not be allowed, and many of the formulas now used must be changed.

Royal's response to Calumet's labels. *Ladies' Home Journal*, Nov. 1922. Author's collection.

*Labels similar to the following will not be permitted:

"This baking powder is composed of the following ingredients and none other: Bicarbonate Soda, Calcium, Acid Phosphate, Anhydrus Soda Alum, Corn Starch."

This should read as follows:

"This baking powder is composed of the following ingredients and none other: Bicarbonate Soda, Phosphate, Alum, Starch."

*In all cases where changes in label are called for manufacturers should submit labels for our inspection. We trust you will give this your immediate attention.[2]

Violations of Minnesota's food and labeling laws were widespread. From January 1, 1905, to August 1, 1906, many categories of food, such as pie filling, relish, and grape juice, were cited for only a handful of violations. Preserves, cider, and milk had approximately two dozen violations each. But liquor led the way, with 179 violations for selling adulterated liquor. Baking powder was second, with 99 violations for selling adulterated or improperly labeled baking powder.[3]

One Minnesota food inspector's report stated the optimism and crusading spirit of the Progressive Era and the hope of moral improvement through legislation: "With a national law with its moral effects on the manufacturers, the time is not far off when illegal foods will be a thing of the past, and it will be a Godsend to humanity." Another food inspector wrote: "I feel that ours is a glorious work for humanity."[4]

Royal and Calumet also fortified their companies through upstream and downstream vertical integration. In 1907 Royal acquired control of Western Glucose. The Hammond, Indiana, company manufactured cornstarch, the critical buffer ingredient in all types of baking powder that prevented the other elements from combining prematurely. In 1908 Ziegler changed the name of Western Glucose to American Maize-Products. The company is still in existence and manufactures high fructose corn syrup.[5]

Calumet, too, integrated vertically both up and downstream. To make Calumet "completely self-sufficient," Warren Wright bought a chemical factory, a canning facility, and a printing press.[6] Both Royal and Calumet frequently published proprietary corporate cookbooks. So did Davis in New Jersey, KC in Chicago, and Rumford in Massachusetts. The recipes were not interchangeable. Without standardization in baking powder, chaos and confusion continued in the cookbooks.

President Roosevelt's Scientific Big Stick: The Remsen Board

In 1908 the baking powder war, as part of the larger controversy over the increase of chemicals in food and drugs, reached President Theodore Roosevelt personally. On the advice of his physician, Roosevelt used the low-calorie sugar substitute saccharin, which some scientists had begun to claim was dangerous. Determined to obtain a definitive scientific ruling, Roosevelt used the power of the presidency to investigate, among other things, the effects on human health and nutrition of saccharin; alum baking powders; sulfur dioxide, a preservative in dried fruit; and salicylic acid, the main ingredient in aspirin.

By executive order Roosevelt appointed a referee board. The illustrious board had five members, some of the top scientists in the United States, all professors. Three were chemists from Yale, Northwestern, and the University of Pennsylvania; one was a pathologist from Harvard. They became known as the Remsen Board after the committee's chair, world-renowned chemist Ira Remsen, MD, PhD, of

Johns Hopkins University.[7] This was like sending the fox to guard the henhouse, as saccharin had been discovered in Remsen's laboratory.[8] However, at the time, this was not perceived as a conflict of interest.

A scientific board of review was not a new concept, but having the president intervene and appoint one was; it indicated an increase in the power of the executive branch. In April 1882 the National Academy of Sciences had appointed five eminent scientists to examine the new product glucose, a sweetener made from grape or cornstarch. The scientists' mandate was to compare glucose to cane sugar and determine if it caused any "deleterious effect" when consumed. That committee had been formed at the request of the commissioner of the Internal Revenue Service, not of the president of the United States, because the presidency then was not the powerful office that it became under Theodore Roosevelt. Two years after the committee was formed, it reported its findings: glucose was "unobjectionable" and was "in no way inferior to cane sugar in healthfulness . . . even when taken in large quantities." Ira Remsen had been one of the five scientists on that board too.[9]

In 1914, after thorough experiments on human ingestion, digestion, and excretion, the Remsen Board issued its findings: aluminum compounds in the doses in which they were consumed in cakes, baking powder biscuits, and other foodstuffs had no toxic effects on humans or food.[10] Like the Pure Food and Drug Act, the Remsen Board findings had no effect on the baking powder war except to serve as more ammunition, another scientific citation.

The formation of the Remsen Board in 1909 was a slap in the face of Harvey Washington Wiley. In 1902 Wiley had conducted what he considered definitive experiments on some of the ingredients the Remsen Board had investigated and had issued his own rulings. The experiments had been highly publicized because he had fed various substances to healthy young men, christened the "Poison Squad" by the press. The 1914 findings of the Remsen Board, many of which contradicted Wiley's findings, were a body blow. But Wiley had already left the USDA. He had departed two years earlier, under a powdery white baking powder cloud of suspicion. Wiley resigned from the USDA on March 15, 1912, after the press printed stories that he gave preferential treatment to a baking powder business connected to Royal; he supposedly "abated" a "case against a cream of tartar baking powder that contained lead."[11]

Wiley had been tilting at some very large windmills with an inadequate lance. Conservative and antimodernist, he had consistently objected to new products that American consumers embraced. Wiley objected to salicylic acid, the main ingredient in aspirin. He objected to sulfur dioxide, which is still used to preserve dried fruits like raisins, apricots, and prunes. In 1911 Wiley was the motive force behind a disastrous lawsuit against Coca-Cola, *The United States v. Forty Barrels and Twenty Kegs of Coca-Cola*. The U.S. government lost.[12]

Wiley Goes to *Good Housekeeping*

> Dr. Wiley has much to say to the women of America, and through the pages of this Magazine, he expects to deliver his message.
>
> —"Special Announcement," *Good Housekeeping*, April 1912

On April 1, 1912, Wiley began a new career at *Good Housekeeping* magazine, where he could continue his crusade against Calumet and the other alum baking powders, but from a different pulpit.[13] In his self-serving books, Wiley says that at *Good Housekeeping* he felt free to "carry on my battle for pure food, finding no enemy to stab me in the back."[14] The enemy was the alum baking powder companies and Wiley's modernist colleagues who did not find them objectionable.

Good Housekeeping, like *Godey's*, was a national magazine for women. In 1890 *Good Housekeeping* had a circulation of four hundred thousand. This was approximately the combined total of the four major magazines at the time: the *Atlantic*, *Harper's*, *Scribner's*, and *Century*.[15] In 1911 William Randolph Hearst purchased *Good Housekeeping* and moved it from Springfield, Massachusetts, to New York City, which had become the center of the new national press.[16]

Before Wiley arrived, *Good Housekeeping* had created a "Tested and Approved" seal. With the advent of electricity in the home and a plethora of new appliances, consumers were at a loss as to how to distinguish among the products. The Good Housekeeping Institute conducted tests and published the results. They debuted the "Tested and Approved" column in the December 1909 issue and approved twenty-one household appliances. With Wiley's arrival, *Good Housekeeping* extended the seal of approval and created the "Tested and Approved Bureau of Foods, Sanitation, and Health Seal" specifically for him.[17] Consumers in the twenty-first century still know this as the *Good Housekeeping* Seal of Approval.

Wiley had extraordinary power at *Good Housekeeping*. It extended to "advisory supervision of the advertisements of foods, remedies and cosmetics admitted to its columns."[18] His columns were very influential in how businesses marketed products and in which ones succeeded. According to Wiley, "Most of the leading manufacturers and advertising men . . . modif[ied] claims, labels, and advertisements when inconsistencies and inaccuracies were brought to their attention"—in other words, when Wiley threatened to withhold approval.[19]

Wiley used a numeric rating system to classify products. Items that rated from 85 to 100 received stars, which indicated "high quality and full weight and measure with accurate labeling and reasonably conservative claims." Products that scored 76–84, inferior in labeling and quality, were rated "N" for "Noncommittal." The lowest rating, below 75, was for products that were harmful or made fantastic claims. These received a "D" for "Disapproved."[20]

Wiley's inaugural column was about the product that he despised and that he believed had brought about his downfall: baking powder. All the baking powders in the Royal Baking Powder Trust received stars. Companies that Wiley sprinkled with stars were then or later became industrial food giants: Campbell, Heinz, Libby, Dole, Nabisco, Kellogg, Postum, Quaker, Ralston, Armour, Crisco, Beechnut (then manufacturing peanut butter), Swift, Hormel, and American Sugar Refining (Domino). Calumet received a "D" rating, as did all alum baking powders.[21] Wiley's "D" rating might as well have stood for "Disappeared." Companies that received a "D," with the notable exception of the baking powders, no longer exist.

"What a Fumin' about Egg Albumen": The Alum versus Alum War

In June 1913, a year after Wiley left the USDA and a year before the Remsen Board issued its findings, a war for control of the alum baking powder market unexpectedly broke out. It was another advertising war of scientific experts, this time between the two most powerful alum baking powder companies, KC and Calumet, both based in Chicago. But it began in the South and was waged primarily in the Midwest and Western states. At the seventh annual convention for the Association of American Dairy, Food, and Drug Officials, in Mobile, Alabama, Jaques Manufacturing Company, which produced KC Baking Powder, began to speak out against the use of egg whites in baking powder. This was aimed at Calumet, which had incorporated egg whites into its formula since its inception. The attack was a shock. Jaques had been firmly allied with Calumet in the 1899–1906 baking powder war against Royal and the Missouri law. Jaques and Calumet had been motivational forces in organizing the American Baking Powder Association, which had represented the manufacturers of alum baking powders against Royal's cream of tartar.

In this new war between the alum baking powder companies, Jaques used the same tactics that Royal had used against the alum baking powders fifteen years earlier. Jaques pressured state legislatures to pass laws making Calumet's use of egg whites illegal. Jaques also used advertising to try to turn the public against Calumet. Jaques even allied itself with Royal against Calumet. Jaques accused Calumet of fraud; Calumet said Jaques was "playing understudy to the trust."[22]

The center of the controversy was Calumet's water glass test. When a teaspoon of Calumet baking powder was stirred into a glass that contained three or four teaspoons of water, the albumen in the egg white foamed up and trapped the gas bubbles that the leavening agents in the baking powder created. It was an impressive visual. Calumet said that the water glass test served as quality control because it provided consumers with "a dependable commercial test" to determine if their baking powder was active or had begun to deteriorate.[23] It was also quality control for Calumet, which claimed that it took back almost eighty-one tons of deteriorated baking powder in one twelve-month period, at a cost of almost twenty-four thousand dollars.[24]

Calumet was disingenuous at best. It was one thing for Calumet to perform the test on batches of its own product in the warehouse, but Calumet's door-to-door sales teams performed the water glass test in front of housewives in their homes. The test worked *only* on baking powders that contained egg white, but Calumet's salespeople compared it to baking powders that did not contain egg white. Other baking powders could have full leavening strength but would fail to foam up because they contained no albumen. It made Calumet look potent and all the others—alum, cream of tartar, and phosphate—look feeble.

Jaques considered this unfair competition and wanted albumen in baking powder declared illegal. In 1913 and 1914 Jaques agents traveled throughout North Dakota, Nebraska, Kansas, Texas, Idaho, Wyoming, Montana, Utah, and Oregon. They advertised, filed lawsuits, and pressured state legislatures to pass laws making albumen and the water glass test illegal.

Although Jaques raised the water glass issue at the state level, Calumet chose to fight this battle at the federal level for several reasons. First and foremost, the

Calumet's water glass test. Calumet claimed that on the left was cheap or "Big Can" baking powder (Jaques); in the middle was cream of tartar baking powder (Royal); and the high-rise foam on the right was Calumet. Government Agricultural Experiment Station, Agricultural College, North Dakota, Aug. 1923. Courtesy of Nach Waxman.

USDA had previously stated that egg albumen was an acceptable ingredient in food, so Calumet was reasonably certain of a favorable outcome. Second, a federal ruling would safeguard Calumet's interstate trade and trump any unfavorable state laws that Jaques might be successful in passing. Calumet and Jaques both filed briefs with the solicitor of the Board of Food and Drug Inspection at the USDA.

By this time, everyone at the USDA was wise to the fact that dealing with baking powder companies was like walking through a minefield. In an internal communication, the solicitor issued a caveat to the chief of the Bureau of Chemistry. No official "Food Inspection Decision should be issued on this subject. The controversy before the Department seems to be a mere trade fight. Every precaution should be taken to avoid the possibility of this Department being used, to advance the interests of either side under such circumstances."[25] On January 17, 1914, the USDA Bureau of Chemistry issued a narrow neutral ruling: "The addition of this amount of pure dried egg albumen to baking powder does not constitute adulteration within the meaning of the [Pure Food and Drug] Act."[26]

However, the ruling did not deter Jaques and Royal from making albumen an issue at the July 1914 National Dairy Convention in Portland, Maine. Although the organizers blocked Jaques and Royal from raising the issue on the floor in the open convention, the two companies did manage to convince convention members to pass a handwritten addendum that condemned the use of the water glass test.[27] Jaques then sent notes to its trade contacts advising them not to use baking powders with albumen: "Mr. Dealer, You have worries enough now without getting into trouble with the Pure Food Authorities. / Baking Powder containing albumen (sometimes called white of egg) has been discredited everywhere."[28]

This advertising was as deceptive and false as the egg white test it decried. Both statements were true, but the second sentence did not follow from the first. Certainly, grocers and jobbers did not want to run counter to the Pure Food Law. But baking powders that contained albumen were not in trouble with the pure food authorities as the statement implied. Nor had they "been discredited everywhere."

Calumet went on the offensive in 1914. It published *Why White of Eggs*, subtitled "*What a Fumin' about Egg Albumen*." The book was 203 pages of scientific evidence and propaganda that aimed to discredit Jaques' accusations. Calumet's chemists claimed that they had found variables and deviations from the scientific method in the tests performed on Calumet baking powder in state laboratories. In Kansas, test ingredients were supposedly not mixed consistently. In Idaho, the mixtures were allowed to rest for different lengths of time and possibly baked at different temperatures.[29] *Why White of Eggs* also reported that federal courts in Idaho and Oregon found nothing objectionable in egg whites in baking powder and that no state had "a law prohibiting or regulating the use of white of eggs."[30] The courts "permanently enjoined the food commissioners from further interference with the sale [of Calumet]."[31]

In August 1914 the *American Food Journal*, a national trade publication headquartered in Chicago, weighed in on the side of Jaques and Royal. The journal devoted its entire front page to "The Albumen Fraud." It reprinted an editorial it had published in June 1914 and again called on all baking powder manufacturers to cease using egg white, because "it serves no purpose but to deceive." The article reported that a federal court in Idaho had declared the water glass test illegal, in spite of what Calumet said. The editorial also exposed how Calumet had manipulated the press and sent out press releases as straight news—in other words, Calumet was now employing the same deceptive advertising tactics that Royal had used for years and for which Calumet had excoriated Royal.

However, in the intervening years, newspaper editors and much of the public had become more sophisticated about press manipulation. Certainly this was the case on the East Coast and in urban areas, although "some of the Washington [DC] correspondents of the Western press have been sending the [fake] stuff out as real news." The *American Food Journal* editorial contained a caveat: "Manufacturers would better realize that the days have gone when it is possible to put over a fraud through the employment of news fakirs."[32]

On February 26, 1918, the USDA issued Food Inspection Decision No. 174, which finally created an official standard for baking powder. The USDA defined baking powder not by its ingredients, but solely in terms of its leavening ability. This was the standard that Calumet and the other alum baking powders had proposed in 1904 at the U.S. Government Committee on Food Standards hearing. The USDA defined baking powder as a "leavening agent produced by the mixing of an acid reacting material and sodium bicarbonate, with or without starch or flour. It yields not less than twelve per cent (12%) of available carbon dioxide." The definition was also approved by the Association of American Dairy, Food, and Drug Officials, where Jaques had started the controversy five years earlier. Although the ruling seemed to standardize ingredients, "acid reacting material" left manufacturers a great deal of leeway in the use of ingredients, because it included cream of tartar, phosphates, and alum. So there was still no standardization of baking powders.

Herman Hulman had not been involved in the alum versus alum war. July 4, 1913, was the first day since his wife, Antonia, had died in 1883 that Herman did not go to the cemetery to visit her. He died peacefully at home on the birthday of his adopted country and was buried beside Antonia.[33] In his will Herman validated the faith that his friend Eugene Debs had placed in him: he gave bonuses to all Hulman employees. Those who had been with the firm less than ten years received two weeks' extra pay; more than ten years, four weeks' extra pay. His son Anton became undisputed head of the firm.[34]

The Biscuit Crusade

In the first decades of the twentieth century, baking powder, along with wheat flour, inadvertently received a boost from medicine and women in the new field of home economics. A rise in the disease pellagra was thought to be caused by musty or moldy corn. The medical profession was divided on the cause of pellagra. Some physicians thought it was infectious, transmitted person-to-person. Other scientists, following in the footsteps of Pasteur and Koch, sought a solution in bacteriology and tried to isolate the organism in the corn that they believed was the cause.[35]

Neither hypothesis was correct. Pellagra was not a communicable disease, and it was not caused by what was in the corn, but what was absent from it. Pellagra was first identified by Italian physicians in the nineteenth century when Italians began widespread consumption of American maize as a staple food called polenta. "Pellagra" is an Italian word that means "rough skin," one of the symptoms of the disease. Without nixtamalization, the addition of vegetable ash, which the Aztecs had used to process corn, it was deficient in the necessary B vitamin niacin. Europeans had taken the grain native to the Americas but not the nourishing technique the Native Americans used to process it.

As the theory that corn was the culprit gained ground, especially in some areas of the American South, the solution was clear: replace corn with wheat, cornbread with wheat bread and biscuits. The home economists' missionary zeal led them to Appalachia, where they tried to replace the traditional corn bread with wheat biscuits. These Northern home economists regarded the residents of Appalachia the same way that Christian missionaries looked at Native Americans: as uncivilized. Like the religious missionaries on the reservations, home economists, too, regarded the ultimate sign of civilization as bread and biscuits made from wheat.[36]

However, the flour that replaced corn was not nutritious dark whole wheat, but refined and white, run at high speed through steel rollers. Lower in gluten than bread flour, it did not bake up elastic and chewy, even in yeast breads. It made a soft crumb that was better for foods made with baking powder: biscuits, cakes, pancakes, waffles, and quick breads. Most importantly, it baked up white.

Beaten Biscuits

In the South, technology also revived beaten biscuits. Because they were so labor-intensive, beaten biscuits had become less common after the end of slavery and the Civil War and therefore even more prized. By the end of the nineteenth century, a machine kneaded the biscuits. It had two rollers, like a washing machine wringer or an Italian pasta machine. Instead of beating the dough with a rolling pin, axe handle, or other implement hundreds of times, the cook passed the dough through the rollers from 150 to 200 times. Some recipes called for a minimum of 300 passes through the machine for family biscuits; for company, 500 passes.[37]

The first recipe in *The Blue Grass Cook Book* (1904), a compilation of recipes by upper-class women in Kentucky, was for beaten biscuits. So was the second. The third entry was "Beaten Biscuit Suggestions." Whether the biscuits were beaten or rolled, the hand that did the beating or cranked the rollers was likely to be African American.[38]

The "Settlement" Cook Book—Milwaukee, Wisconsin

At the same time that baking powder was being forced on people in the rural South, people in the urban Midwest were also being indoctrinated in its use. Like women in Kentucky who had compiled *The Blue Grass Cook Book*, women in Milwaukee, Wisconsin, had compiled *The "Settlement" Cook Book* in 1901. Settlement houses were created throughout the United States to help immigrants "settle" into their new country. However, *The "Settlement" Cook Book* is unique because it was compiled by German Jewish women to assist the more recent Eastern European Jewish immigrants. The ostensible reason for creating the book was to aid students and to raise money for the settlement house, but the underlying reason was that the earlier German Jews considered the Eastern European Jews to be of a lower class, which they feared would cause a rise in anti-Semitism.

The book's main compiler was Mrs. Simon Kander, who was aided by three other women. The book's subtitle explains some of its sources: *Containing Many Recipes used in The "Settlement" Cooking classes, the Milwaukee Public School Cooking Centers, and gathered from various other Reliable Sources.* As with the cookbooks of other immigrant groups, baking powder appears frequently and prominently, especially in culturally significant foods they made often or for holidays. It is also indicative of how widespread this new American technology had spread and how much the Jewish population had assimilated. After yeast, baking powder was the preferred method of leavening, followed by baking soda and vigorous egg beating.[39] *The "Settlement" Cook Book* did not even entertain the idea of alum chemical leaveners. It stated unequivocally: "Baking Powder is composed of baking soda and cream of tartar, with a little flour or cornstarch."[40]

Baking powder–leavened biscuits, breads, cakes, cookies, and rolls are included in several chapters throughout the book. Almost every recipe in "Biscuits and Breakfast Cakes," contains baking powder. First are the ubiquitous baking powder biscuits, followed by American standards like strawberry shortcake (two recipes), and American-style pancakes, called griddle cakes (six different kinds). In a separate chapter called "Eggs, Omelets, and Pancakes" are recipes for French pancakes and German pancakes, which contain no baking powder, are thin, and cooked until crisp—essentially, crêpes. Potato pancakes, a staple Jewish American holiday food, contain "A pinch of baking powder."[41] Even the "Entrees" chapter begins with "Batter for Fritters" made with baking powder.[42]

The chapter "Bread, Rolls and Toast" contains quick breads like currant bread and corn bread, leavened with baking powder, along with yeast-leavened breads. In "Dumplings and Garnishes for Soup," dumplings, cooked by dropping raw dough onto soup, are made with wheat flour or grated almonds ("Mandel Kloese"), but both are leavened with baking powder.[43] The chapter titled "Mehlspeise—(Flour Foods)," includes more recipes for dumplings, several leavened with baking powder, and which precede yeast dumplings.[44] Pie dough, short crust pastry, and doughnuts contain baking powder.[45] Even puddings contain baking powder: "Suet Pudding No. 2," "Steamed Fruit Pudding," "Chocolate Pudding No. 1," and "Steamed Chocolate Pudding."[46]

Baking powder is prevalent in cakes as well. "Sponge cakes contain no butter and are made rich with eggs; the lightness depends upon the amount of air beaten into the eggs."[47] Yet every sponge cake recipe, except for those containing matzo, which would have remained unleavened for religious purposes, contains a chemical leavener. Most use baking powder or its equivalent, cream of tartar and soda, or cream of tartar or soda by themselves. Even the traditional anise-scented zwieback (twice baked), like biscotti, contain baking powder, as do both chocolate layer cake recipes.[48] Likewise, although the section "Torten" claims that these are "cakes that contain no butter, but are made . . . light with eggs," they are also made light with baking powder. Baking powder provides the leavening in more than a dozen tortes, including "Chocolate Zweiback Torte," "Poppyseed Torte," and "Rye Bread Torte."[49]

Small cakes and cookies also contain baking powder, and reveal their European origins and American assimilation: "Anise Cookies (Springerle)," "Pfeffernuesse," "Gingerbread," and "Lebkuchen No. 2" for the former; and "Peanut Drop Cakes," "Chocolate Cookies," and "Oatmeal Cookies" for the latter.[50]

Cookies and Christmas Dollies

The many cookie recipes in *The "Settlement" Cook Book* reflect their rising popularity in the beginning of the twentieth century. The word "cookie" had appeared for the first time in Amelia Simmons's *American Cookery* in 1796; they were some of the first recipes that contained pearlash. As William Woys Weaver has pointed out, those cookies, made in the home, were round because the technology did not exist to make them into other shapes. The dough was either dropped onto sheets and baked, or rolled and cut out with the rim of a drinking glass, wine glass, teacup, or canister lid. During the 1880s, however, tools like tin cookie cutters, previously available only to professional pastry cooks and confectioners, were mass produced and affordable. Industrialization brought them into American kitchens, with an added aid to the cook: the cookie cutter imprinted lines onto the cookie to show where the icing should go.[51]

Among the most common cut and decorated cookies were "Yule Dollies." These were a combination of two disparate traditions. One was the medieval European Christmas tradition, especially in Germany, of pastries shaped or stamped into figures or scenes, like gingerbread people or springerle. German immigrants to America often brought molds made by master carvers with them.[52] The other tradition was from the early nineteenth century, when Wales promoted tourism through quaint costumes with tall hats.[53]

YULE DOLLIES

(from Recipe of Cornelia C. Bedford, *Table Talk 14* [November 1899], 435)

8 Tablespoons (1 stick) unsalted butter
1 cup sugar
2 eggs
1 Tablespoon heavy cream
1 teaspoon vanilla extract
3 cups all purpose flour
2 teaspoons baking powder

Cream the butter and sugar. Beat the eggs to a froth and combine with the sugar mixture. Add the cream and vanilla. Sift the flour and baking powder together twice, then sift into the batter. Work this into a dough, cover, and rest it in the refrigerator for 1 hour before baking.

Preheat the oven to 350 F.

Roll out the dough to about ½ inch thick and cut out the dollies with a tin cutter. Set them on greased baking sheets and bake in the preheated oven for 15 minutes. Cool on racks.[54]

The "Yule Dollie" was then decorated with royal icing made from confectioners' sugar, egg whites, lemon juice, and various food colorings.

While the dollies are a variation on shortbread cookies, the 208 cookie recipes that a woman named Anna Covington collected from 1917 to 1920 are an important step in the evolution of the cookie. In 2005 writer Barbara Swell arranged and annotated the recipes into *The 1st American Cookie Lady: Recipes from a 1917 Cookie Diary*.[55] The American predilection for molasses, sugar, and ginger cookies remains strong, as it was in Amelia Simmons's first cookie recipes. Some of Anna's cookies are still small cakes, but almost all are leavened with baking powder or baking soda. Only 7 of the 208 recipes are for what Anna called chocolate cookies. Three of those are brownies, a new recipe first printed in 1906 in the *Boston Cooking-School Cook Book*.[56] The chocolate in these recipes is unsweetened and melted, cocoa or grated. Missing are chocolate chip cookies, still two decades away.

The recipes were handwritten in the practiced penmanship of the time. This is reminiscent of the heirloom manuscripts of Martha Washington's time. It is also an anachronism. By World War I, typewriters had replaced handwriting. Home economists were turning to recipes that were typed onto 3" × 5" index cards that could then be alphabetized and used one at a time instead of having to search through a book. If they got spattered with batter, they could be replaced instead of ruining a book or getting the pages stuck together.

Many of Anna's cookie recipes came from magazines such as the *Ladies Home Journal* and *The Modern Priscilla* and from Royal Baking Powder cookbooks.[57] Her compilation reveals the chaos that still reigned in cookbooks because of baking powder. Anna has several different recipes for the same kind of cookie but with different amounts of baking powder. For example, two recipes for sugar cookies are similar except that one uses two teaspoons of baking powder and the other uses three.[58] However, she makes no comment on the differences in baking powder.

World War I: Doughboys and Doughnut Girls

America's entrance into World War I in April 1917 created opportunities for baking powder manufacturers on the home front and on the front lines. Refined white flour was limited on the home front, so Americans were forced into being gluten-free or nearly so. The Federal Food Administration, set up to deal with food during the war, used women's magazines to urge America's "culinary soldiers" to sign pledge cards and mail them to Washington, D.C., to Herbert Hoover, head of the Food Administration. The pledge was to plan one wheatless and one meatless meal every day, observe wheatless Mondays and Wednesdays, meatless Tuesdays, and no-pork Saturdays.[59]

Housewives made War Bread and Victory Bread out of unrationed corn, rye, oats, rice, and barley flour. All contained little or no gluten, so yeast could not leaven them; only baking powder or baking soda could. Sugar, too, was in short supply, so recipes used substitutes like molasses, honey, or corn syrup. The substitutions made passable cakes, cookies, and other baked goods, but the recipes generally were retired when the war ended.

RYE DROP CAKES

1 heaping cup rye meal, 1 heaping cup flour, pinch salt, 3 tablespoons molasses, 1 cup milk, 2 eggs, 2⅓ teaspoons baking powder. Sift dry materials together. Add milk, molasses, & beaten eggs. Drop by spoonfuls.

BARLEY & ROLLED OAT DROP COOKIES

Cream together: 1 cup fat, ¼ cup brown sugar, add ½ cup corn syrup, 1 egg, beaten. Mix 1 cup barley flour, 2 teaspoons baking powder, pinch salt, ½ teaspoon cinnamon. Add to the first & mix well & then add ½ cup nutmeats & 1 cup raisins. Drop from teaspoon on well greased tins & bake in moderate oven.[60]

Eggs, too, were in short supply and expensive for home use. In 1917 Royal published *55 Ways to Save Eggs*. The twenty-two-page cookbooklet introduced a "New Way" of baking that promised savings with no reduction in quality over the "Old Way": simply substitute one teaspoon of Royal baking powder for each egg omitted. The booklet encouraged homemakers: "Food made at home is not only more economical and of better quality, but will keep fresh longer. Also there is the added advantage of knowing that the ingredients used are healthful." The New Way and the Old Way were presented side by side. One of the recipes was for "Eggless, Milkless, Butterless Cake," which called for five teaspoons of baking powder. The recipe was resurrected during World War II, when it became known simply as "The War Cake."[61]

Wheat and sugar were available for the military, however, and made their way to the fighting men in a form that became popular during the war and even more popular in the United States after: doughnuts. Salvation Army women wanted to bring a taste of home to the doughboys. Their first thought was to supply slices of pie with the cups of coffee they handed out. But conditions at the front were rough, and pies require preparation: clean water to wash the fruit for the filling; knives to peel and slice the fruit; rolling pins and a clean surface to roll out the dough; pie pans, and ovens in which to bake the pies. Flaky pie crust requires skill and can be difficult to make. Like yeast, it is sensitive to weather, especially humidity. So are European-style doughnuts, such as round, jelly-filled German doughnuts; French beignets; Italian *zeppole*; Pennsylvania Dutch *fastnachts*; and Dutch *olie-koecken* (literally, "oil cakes"). They are all yeast-risen and therefore subject to the same vagaries of temperature, humidity, and strength of yeast that can make bread baking difficult.

American doughnuts, on the other hand, are made with baking powder. These "cake doughnuts" require less skill, few ingredients, and minimal equipment. Flour, sugar, baking powder, canned milk, shortening, a bowl and a spoon for mixing and dropping the dough, and oil for frying are all that are necessary. American doughnuts require no advance preparation and no time for yeast to rise. Because they are leavened with baking powder, they are not affected by temperature or humidity. Waiting for the oil to heat up so the doughnuts can be fried is the longest part of the preparation time. If time and space permitted, the doughnuts could be rolled out and cut or shaped.

The men appreciated the Salvation Army's hot coffee and warm doughnuts, consuming as many as nine thousand a day. The women who made and served the doughnuts became known as "Doughnut Girls." Other charitable organizations like the YMCA and the Red Cross followed the Salvation Army's lead.[62] Margaret Sheldon, one of the "Sallies," as the Salvation Army women were called, estimated that she made one million doughnuts during the war.

SALVATION ARMY DOUGHNUTS

Makes 400 large doughnuts or 500 small doughnuts

18 pounds of flour
7 pounds sugar
12 ounces of good baking powder
3 ounces salt
3 ounces of good mace
6 big cans of evaporated milk
8 cans water
1 pound lard[63]

Misbranded: Outlaws in Montana

After a brief patriotic lull during World War I, the baking powder war between Royal and Calumet resumed on all fronts—advertising, state, and federal. The Royal Baking Powder Trust found itself afoul of the law again. Many states had passed pure food laws patterned after the federal Pure Food and Drug Act. This allowed them to take action locally and quickly without having to rely on or wait for the federal government. One case occurred in Montana.

In 1919, after nearly sixty years as a cream of tartar baking powder, Royal changed the Dr. Price formula. It removed the cream of tartar and replaced it with the cheaper mineral phosphate. This placed Dr. Price's Cream of Tartar in the same category of baking powder as Rumford, which Royal had been claiming was adulterated since *The Royal Baker and Pastry Cook* in the 1870s. It also allowed Royal to reduce the price by half. But Royal, which had clamored for truth in labeling, did not change the name of the baking powder and made only minor changes to the label, such as putting quotation marks around the word "Cream." Royal also neglected to announce, except to the trade, that it had changed the formula.

In 1920 the Montana State Board of Health sent letters to wholesale and retail grocers stating that it would contest the sale of Dr. Price's "Cream" Baking Powder on the grounds that it was misbranded. Royal obtained a temporary injunction to prevent the Montana Board of Health from stopping sales.

On May 12, 1920, the district court ruled strongly in favor of the Montana State Board of Health and against Royal. The judge's tone in the written decision was vexation at Royal on multiple counts. One was simply the sheer volume of Royal's "very elaborate briefs and arguments, which range unduly wide." Another was the sudden shift in formula, because "the superiority of baking powder containing tartars from grapes as the acid ingredient, over those containing phosphate or alum . . . was diligently impressed upon the public." The court also stated that regarding the words and artwork on the label, the public "are habited to its general

The New Dr. Price Cook Book, 1921, after the Montana court ruled regarding deceptive advertising, and the formula changed to phosphates. Author's collection.

appearance and accept it." Further, the burden was never on the buyer, but always on the seller to ensure that the label was clear: the law "lays no command on the purchaser to scrupulously or at all read labels." Later in the same paragraph: "Vendors are not permitted to entrap the ordinary careless and ignorant." In its summation, the court precluded any further action on the part of Royal: the board of health's "judgment is final and not to be reviewed and set aside by any court. The intent of complainant is immaterial. Its belief is immaterial. The only material matter is its actual failure to measure up to the requirements of the law, in defendant's judgment, in the matter of the label."[64]

World War I was over, but the baking powder wars continued. As the 1920s began, the baking powder wars entered their sixth decade. Battles on all fronts were coming up: newspapers, magazines, courts, state, and a new federal agency.

CHAPTER 9

The Federal Trade Commission Wars

The Final Federal Battle, 1920–1929

Alum is a mineral substance . . . largely used for kitchen utensils.
—Royal anti-alum baking powder advertisement, 1920s

Tartrates injure the kidney causing nephritis.
—Calumet anti–cream of tartar baking powder advertisement, 1920s

By the 1920s the public was reaping the benefits of the Industrial Revolution, which had reached America's home kitchens. Electricity, indoor plumbing, refrigerators, sinks, and stoves revolutionized cooking and baking. By 1915, oven thermostats were nearly universal. Smaller appliances like blenders and toasters made cooking easier. The electric mixer replaced the rotary hand mixer. In spite of new appliances, before the discovery of vitamins, wheat-based breadstuffs still remained central to the American diet. A 1922 Metropolitan Life Insurance Company cooking pamphlet advised that "a man of average size, who is moderately active, is likely to be well fed on a diet which includes the following":

16 ounces of bread or cereal
16 ounces of milk
8 ounces of meat, fish, eggs, cheese, or legumes
5.3 ounces of potatoes or root vegetables
5.3 ounces of vegetables and fruit
2–3 ounces of sugar
2 ounces of fat[1]

The first section of the book was nine pages of breadstuffs leavened with baking powder: griddle cakes, waffles, muffins, biscuits, tea cake, short cake, coffee cake, and "Baking Powder Breads." It was followed by three pages of yeast-risen bread and rolls.

The year before, 1921, had been the best year in Royal's history. An article in the *American Food Journal* pointed out how global Royal had become: "Royal Baking Powder was used by the Peary Expedition to the North Pole, and by the Scott expedition to the South Pole. Its empty tins have been found in the interior of Africa, on the slopes of the Andes, in the prospectors' camps of Alaska and on the desert of the Sahara."[2]

However, in the 1920s the rise of commercial bakeries caused a "decline in home baking and a tremendous decline in the use of baking powder."[3] The professional baking industry embarked on aggressive campaigns to sell more bread. There was the "Eat More Wheat" campaign, followed by the "Eat More Toast" campaign. The decline in home baking created even more competition in the baking powder industry. Baking powder companies, like other companies, created test kitchens staffed by female home economists. They also hired well-known women such as Sarah Tyson Rorer, Fannie Farmer, and Mary Lincoln, usually connected to cooking schools, to provide testimonials and to write cookbooks.[4] Throughout the 1920s, Calumet, Royal, Rumford, Cleveland, and others continued to produce promotional cookbooks. Calumet published more titles in its Reliable Recipes series, along with other titles like *The Master Cake Baker*. Royal's education division produced *Anyone Can Bake*.

In the 1920s Royal and Calumet continued the baking powder war at the state level and escalated it to a new battlefield and a new agency at the federal level, the Federal Trade Commission (FTC). The FTC was five years old when Calumet filed a protest regarding Royal's advertising. On February 4, 1920, the FTC filed a complaint against Royal because Royal "had disparaged and defamed goods of its competitors and had falsely charged competitors' baking powders as being poisonous." Royal had also claimed that alum baking powders were "made from ground-up aluminum cooking utensils."[5] The FTC sought to prevent Royal "from making any reference whatever to the presence or absence of alum in its own product or in the products of its competitors," because "it would be in the interests of the public."[6] Royal responded to the FTC's order to cease and desist its advertising by ceasing and desisting. As far as the trial examiner was concerned, that was the end of the case. He issued an order of dismissal.

However, Royal had been doing some investigating of its own. Calumet was shocked to discover that four of its top salesmen were actually spies for Royal. In 1923 Royal had hired the William J. Burns International Detective Agency to infiltrate Calumet. The agency sent in four operatives undercover as salesmen to gather information that Royal could use against Calumet in its FTC suit. Calumet accused Royal of industrial espionage.[7]

Royal had ceased and desisted, but only temporarily. When the FTC learned that Royal had resumed its use of injurious advertising, the commission vacated its dismissal and reinstated the charges.[8] Royal promptly sued the FTC on the grounds that the Federal Trade Commission Act prohibited the FTC from vacating its own decisions. Therefore, the FTC had no right to reopen a case after they had dismissed it.[9] Many businesses had brought lawsuits against the commission to attempt to overturn its rulings. The courts had overturned FTC rulings in twenty-two cases since the FTC had been created. *Royal Baking Powder Co. v. Federal Trade Commission* was not one of them. Royal joined thirteen businesses that were unsuccessful in their suits against the FTC. The case went to trial.

Royal's trial brief was an exercise in chutzpah. Like the man who kills his parents and then begs for mercy because he is an orphan, Royal claimed that alum was so dangerous that states had passed legislation to protect the public from it. Royal then cited the Missouri law it had bought with bribery.[10] Calumet presented documents from its scientific experts that proved that alum baking powders *were not* injurious to health. In rebuttal, Royal presented its scientific experts' reports, which proved that alum baking powders *were* injurious to health. The testimony of a total of 156 witnesses filled 4,711 pages and contained 632 exhibits.[11]

The whirlwind of absolutely certain and completely contradictory testimony by scientific experts for both sides flummoxed the FTC commissioner. On March 23, 1926, he issued an order of dismissal, which stated: "It may be that, until science has advanced farther, it will not be possible to establish beyond a reasonable doubt whether aluminum compounds as used in baking powders are harmful or are harmless."[12] For a while the baking powder war was at a draw.

The New American Clabber Girl

While Calumet, Royal, and the FTC were embroiled in battle in Washington, DC, Hulman had its own battles in Indiana. The post–World War I labor unrest of 1919, which had seen 20 percent of the workers in the United States involved in strikes, continued into 1920. In January, Hulman granted a selective 8 percent increase: only for workers who made twenty-five dollars per week or less and who had been with the company at least five years. However, there were two major sticking points. Hulman employees were still working a twelve-hour day, and workers objected to the moral judgments and intrusion into their personal lives, because being in arrears on boardinghouse or other bills was grounds for dismissal. On May Day 1920, Hulman & Co.'s workers struck. Hulman declared an open shop.[13] By May 4 the strike was over. The workers received a raise but still did not get the eight-hour day they were demanding. However, Hulman was the only house in Terre Haute that recognized the teamsters and handlers and packers unions.[14]

Herman Jr. was in failing health. In constant pain and wheelchair-bound, he told his brother, "I have two invitations, Anton. You folks have all invited me to stay with you here, and God has invited me home with him." On June 24, 1922, Herman accepted the second invitation.[15]

The following year, on August 17, 1923, Anton filed a name change with the FTC to change "Clabber" to "Clabber Girl." This was to avoid any possible charges by the FTC of misrepresentation or fraudulent advertising. Although baking powder produced results that were similar to the effects of clabbered or naturally soured milk, the chemical processes were different. The name "Clabber" might imply that they were identical or that the baking powder contained milk in dried form, a misconception that the label, with its butter churn, might further. The label was changed to show a young woman sweeping instead of churning. After the Jaques–Calumet battle over albumen, the name change would also prevent attacks from other baking powder companies.[16] The name change went into effect on March 18, 1924.[17]

By renaming the product, Hulman also capitalized on customer recognition of the young woman on the label and on the changing status of women in the United States. The turn of the twentieth century was a time that glorified the "New American Woman." The term was coined in 1894 to describe a newer, freer, self-created woman who was the counterpart and companion to the self-made man.[18] She was college-educated or at least not afraid of using her mind; she worked outside the home, enjoyed public amusements, rode bicycles, and played tennis. These new women were young, slender, and fashionable.[19]

And they were tremendous sales assets as well. "What is the psychology of using a pretty face?" asked a 1902 *Cosmopolitan* magazine article in one of the first instances of the connection of psychology to advertising. Conclusion: the product the young woman was using, in this case a Kodak camera, was "of a highly superior order of merit because of the beauty of face and raiment."[20] There was a rush to feature pretty young women in advertisements during the first two decades of the twentieth century: the athletic Gibson girl, the Coca-Cola girl, the Remington girl (named after the typewriter she used in the office), the Sun Maid Raisin girl, the Land O'Lakes Butter Native American maiden, the Morton Salt girl, the Jell-O girl, Kellogg's Sweetheart of the Corn girl, the St. Pauli (beer) girl, and the Clabber girl.

The Hulmans made another crucial change to the label. The original label showed a pretty young woman churning butter, but on the new label she was carrying a plate of biscuits. In a forest of red and gold high-contrast baking powder labels, the Clabber Girl label was soothing and low-contrast. The background was a demure white, with a gray-tinted scene, almost like a charcoal sketch.

Both the name and the picture on the can harkened back to simpler, homier times, far from the chemistry, urbanization, and technology that had created the

The Clabber Girl. Little was altered from 1923.

can and its contents. The label depicted a capacious family kitchen during the preparations for a holiday meal. In an inglenook in the background, the mother sits in a chair, surrounded by three young children. Of the five people on the label, only one is male, a little boy. The mother plucks feathers from a goose, the traditional centerpiece for German holiday meals. On the floor next to a small toy horse, a cat playfully bats the goose feathers. Perhaps they will be the stuffing for a luxurious pillow.

In the foreground, larger than the other characters, is the Clabber Girl. She represents and appeals to the person in the family who did the baking: the teenage girl. The Clabber Girl is idealized daughter and big sister. She is pretty but not intimidating. She is sweet, innocent, and pure. She smiles as she holds up a plate of biscuits, beckoning you, the honored guest, to join her and the rest of the family at the table. The biscuits have been carefully rolled and cut, not sloppily dropped from a spoon. Dropped biscuits are misshapen and not uniform, so the Clabber Girl is skilled and has taken extra time to make the biscuits look appealing. She is proud of her handiwork and has piled the biscuits into a neat pyramid of abundance. The biscuits also represent what was then the fairly new "pleasure principle" in advertising: food so enticing that it makes the viewer's mouth water in a Pavlovian response.[21]

The Clabber Girl's hair is bobbed, so she is a modern young woman. No long Victorian hair for her. It is also waved, but it is not a Marcel wave; it is a finger wave. Both styles were popularized by Hollywood starlets in the 1920s, but they are very different. A Marcel wave is rigid and industrial. Heat applied through a curling iron forces the hair into tight rows of waves, almost like curls. For a finger wave, as the name implies, the hair is pressed into gentle ridges between the fingers—the same skill that is used in pie making to create the crimp in crust. A finger wave means that the Clabber Girl is modern, but she is not a frivolous painted flapper.

The Clabber Girl also echoes one of the most famous advertising icons in American history. La Belle Chocolatière—The Beautiful Chocolate Girl—had been the unofficial symbol of Baker's Chocolate in Dorchester, Massachusetts, since 1877 and its official trademark since 1883. It was copied from a painting done by Swiss artist Jean-Étienne Liotard in 1745.[22] La Belle is an eighteenth-century serving girl who is carrying a tray with hot chocolate on it for her master or mistress. Like La Belle, the Clabber Girl is crossing from left to right and holds the food in front of her on a tray. However, in the self-conscious age of movies, the Clabber Girl is aware that she is being observed and is comfortable with it. She looks out of the label directly at the viewer. She is self-confident and used to interacting with the world; unlike the European La Belle, this American Clabber Girl is no servant.

Anton Hulman continued to experiment and improve Clabber Girl baking powder, even though it was a local product, sold only in Indiana, Ohio, and Illinois on a small scale.[23] An obstacle to increasing output was that filling, labeling, and packaging the cans was not mechanized. Performing all of this labor by hand was tedious. An inventor was found to create a modern filling machine, which was made to Hulman's specifications in St. Louis.[24]

The Clabber Girl was a teenager, but other baking powder companies used connections to American advertising icons and toys to market their baking powders to younger children. In addition to their original logos, Calumet and Royal also featured characters that would appeal to children on the covers of their cookbooks. Calumet had an Indian chief on its label but a Kewpie doll on its cookbooks. The Kewpie doll, nicknamed for Cupid, was a mischievous character with one large spit curl descending down the middle of his forehead. Created in 1909 by Missouri artist Rose O'Neill, by 1914 the Kewpie doll was the best-selling toy in the United States. The Kewpie doll resembled the Campbell Soup Kids, created in 1905 by another female artist, Philadelphian Grace Gebbie Drayton, and the *New York Journal* newspaper comic strip Katzenjammer Kids, created in 1897 by Rudolph Dirks, a German immigrant.[25]

Calumet's advertising also bragged that it was "The World's Greatest Baking Powder" and the world's best-selling baking powder: "SALES 2½ TIMES THOSE OF ANY OTHER BRAND." On the label on the can, Calumet stated, "NOT MADE BY A

Calumet's mischievous Kewpie doll icon, with his dog. Kraft Heinz Company.

TRUST," in capital letters above the word "Calumet." This was a slight change from "NOT MADE BY THE TRUST," which it had used at least through World War I.[26]

Around 1923, Calumet began to use the slogan that redefined baking powder and with which Calumet became most identified. In upper- and lowercase letters, at the bottom right of its cookbook covers, Calumet added the words "Double-Acting." It made Calumet sound twice as strong as all other baking powders, even though it was not. The double action was first baking soda, then acid. The baking soda in all baking powders reacts with the liquid in the bowl. Bakers call this "bench rise" because the chemical reaction occurs on the bench or preparation table. The second rise is separate, when the acid in baking powder—in Calumet's

Calumet's "Double Acting" advertising. Kraft Heinz Company.

case, the sodium aluminum sulfate—interacts with heat. Bakers call this "oven rise." In addition to bench rise, Wright's new baking powder also had a strong visible oven rise, the kind people still like to peek at through the oven door as cakes bake.

"Double-Acting" was a claim that Royal and the other cream of tartar baking powders could not counter, because cream of tartar is an acid that has bench rise. Everything happens in the bowl, nothing in the oven. As consumers became accustomed to a double-acting, two-fisted baking powder as the norm, cream of tartar baking powders began to look feeble.

In the 1920s Royal countered Calumet's Kewpie doll cookbooks with a series of four cookbooklets for boys. From fifteen to twenty pages each, all of the cookbooklets related, sometimes in verse, how a Royal Baking Powder Trust product solved a problem and made children happy. *Billy in Bunbury, The Little Gingerbread Man, The Comical Adventures of Captain Cooky,* and *The Prince of the Gelatin Isles* were adventure stories that appealed to boys. There were strong male figures in the booklets: kings, chefs, pilots, explorers. Beautifully illustrated by Ruth Plumly Thompson, who had illustrated *The Wizard of Oz* and the other Oz books, the booklets featured recipes for muffins, gingerbread, cinnamon buns, doughnuts, and cookies. The recipes were foods that would appeal to children, not foods that could be made by children, especially doughnuts, which require deep frying. The finale was always a birthday cake.

The plot of *Billy in Bunbury,* which advertises Dr. Price's Baking Powder, is a reworking of *Little Lord Fauntleroy,* a Cinderella story for boys. Serialized in *St. Nicholas* children's magazine from November 1885 to October 1886, and then published as a book in 1886, *Fauntleroy* became a huge hit and set fashions for children for the next fifty years. The boy has hair with curls and wears white collar and cuffs and black knee pants.

The influence of Fauntleroy fashion is apparent in the drawings in *Billy in Bunbury.* Billy is accompanied by King Hun Bun's dog, Ginger Snaps. The Fauntleroy-style boy and his dog were also the basis for a popular comic strip that featured Buster Brown and his dog, Tige (for Tiger), who were also used to advertise Buster Brown bread.[27]

Fauntleroy is about a poor fatherless boy in New York City who discovers that he is really British royalty. The finale is Little Lord Fauntleroy's eighth birthday party, a grand celebration on the castle grounds: "All the tenantry were invited, and there were to be feasting and dancing and games in the park, and bonfires and fire-works in the evening. 'Just like the Fourth of July!' said Lord Fauntleroy." There was a "grand collation," but no specifics about the food.[28]

But to Americans, birthday means cake. In *Billy in Bunbury,* the happy ending on the last page of the book includes a recipe for birthday cake. Candles—the indoor equivalent of fireworks—are placed around the outside of the cake so that it resembles a crown. By the 1920s, Americans were also singing a song written by two sisters in Kentucky for their kindergarten class, "Happy Birthday to You," now the most popular song in the world.[29]

Billy in Bunbury was blatantly anti-mother. Unlike Fauntleroy, who is unhappy because he and his mother are poor, Billy is miserable and skinny because his mother is bad. She does not understand what he likes to eat, which is desserts made with baking powder, especially cake. *Billy* allied the child with the baking powder company in a common goal. It undercut the mother's authority and cast her in the light of incompetence: Your mother doesn't know what you like to eat, so we will tell her. Not only does she not know what little boys like to eat, but she does not have the skill to bake it, so we will show her. Dr. Price's baking powder then becomes the fairy godmother with a yellow can for a wand and solves all problems.

Aping the aristocracy was also evident in new wedding fashions. The March 21, 1871, wedding of a real princess, Queen Victoria's sixth child, Princess Louise, to John Campbell caused a sensation. It was the first wedding between royalty and a commoner since 1515. The cake was multitiered and ornately decorated, what we know now as a wedding cake. This was a huge change from earlier American wedding cakes.

In the eighteenth century and early nineteenth century, American wedding cakes were yeast-risen, bread-like cakes similar to the British great cakes or the American "Election Cake" Martha Ballard made. These fell out of fashion in antebellum America, when pound cakes, which used no yeast, became popular. Both of those types of cakes were dense. They were also yellow because of the eggs and butter. At the end of the nineteenth century, new cakes, made with refined white flour and leavened with white baking powder, were white inside as well as out. The interior purity of the cake symbolized the bride's virginity. The bride looked like a walking confection wearing a veil like spun sugar. Cake and bride were both adorned with flowers—fresh for the bride, fondant for the cake. The flowers often were orange blossoms because of their connection to fertility.

Wedding cakes would not experience another seismic shift until the beginning of the twenty-first century. Then, although wedding cakes were still white on the outside, they could be any kind of cake on the inside: chocolate, carrot, red velvet. The cake itself changed from the traditional progressively smaller multi-tier, to individual cupcakes, or even cakes made from stacked wheels of cheese.

Clabber Girl's War with the Ku Klux Klan

Clabber Girl also had to face a threat that other baking powder companies did not: the Ku Klux Klan. The KKK's goal, as Sinclair Lewis put it in *Elmer Gantry*, his 1927 novel about a hypocritical Midwestern minister, was "to keep all foreigners, Jews, Catholics, and negroes in their place, which was no place at all."[30] A chance meeting with a Protestant minister at a hotel in Terre Haute in 1922 had provided the genesis for *Elmer Gantry*.[31] On Tuesday, April 10, 1923, at 7:30 p.m., the Klan marched from Eighteenth Street west on Wabash Avenue to where it dead-ended at the courthouse by the Wabash River.[32] This route took the Klan directly in front of Hulman & Co.

German communities throughout the United States were still reeling from severe anti-German sentiment in World War I. The xenophobia continued into the 1920s. As Germans and Catholics, the Hulman family personally, and their businesses, were at risk. Although by the 1920s they were second and third generation, they were deeply involved in both the German and the Catholic communities in Terre Haute. They still employed immigrants from Germany and helped them to assimilate. The Hulmans also continued to donate money, land, and time to Catholic churches, hospitals, charities, and other organizations.

The second Klan began its revival throughout the United States on March 3, 1915, bolstered by director D. W. Griffith's profoundly racist epic 190-minute film, *The Birth of a Nation*, in which the Klan cavalry rides to the rescue of the South and white womanhood.[33] By the 1920s Indiana had "the most powerful state Klan in the country."[34] In the 1924 elections the Klan swept the state: governor of Indiana, mayor of Indianapolis and other major cities, and many members of the legislature.[35] Governor Edward Jackson used his power of appointment to give state jobs to Klan sympathizers.[36]

In 1925 Terre Haute had a population of sixty-eight thousand and one of the largest Klan memberships in the state of Indiana.[37] More than seven thousand men, almost one-third of the native white men in Vigo County, were members of Klan No. 7, "an organization that was enormous by community standards."[38] However, the Klan came into violent conflict with the other two-thirds of the population. In 1924 two Terre Haute Klansmen were murdered and Vigo County Klan headquarters were burned down in an arson fire.[39] The *Fiery Cross*, the Klan newspaper, reported that one murderer had stolen a car in "Hunkie Town" and abandoned it in "Grasselli," both identified as settlements of "foreign-born."[40]

The Indiana Klan's power did not come solely from its men. The New American Woman also belonged to the Klan. The Women's Ku Klux Klan (WKKK) was a powerful ancillary organization. Almost one-third of the white native-born women in Indiana belonged to the WKKK. There was a chapter in Vigo County. As many as fourteen hundred women attended events such as a women's minstrel show and an Aunt Jemima Glee Club at the Terre Haute fairgrounds.[41]

WKKK women exercised the power of the purse to support the businesses of "100 percenters"—that is, Klan members—in two ways: first, by patronizing them; second, by driving Indiana's longtime Jewish, Catholic, black, and "alien," or immigrant, population out of business and out of town. The WKKK worked its way through the lists of Catholic businesses and businessmen published in the *Fiery Cross*. The women waged a poisonous whispering campaign of gossip, rumor mongering, and boycotting. One WKKK leader bragged that "she could spread any gossip across the state in twelve hours."[42] In pre-internet days that was wildfire. Members of the Junior KKK for teenage boys and the Tri-K Klub for teenage girls aided their parents in economic terrorism through chapters in fifteen states, primarily in the Midwest. Women of the WKKK also gathered at tea parties, where they served cookies and Kool-Aid and introduced their younger children and infants to the Klan.[43]

On Friday, March 28, 1924, the Ku Klux Klan targeted Terre Haute. "TERRE HAUTE IS SHOCKED" screamed the headline above the fold in the *Fiery Cross*. This was alarming, because not much shocked Terre Haute, which had a reputation as a center for vice. In 1916 there were four hundred madams in operation and nine hundred prostitutes.[44] The *Fiery Cross* article stated that the sheriff had corrupted all of Vigo County by not enforcing Prohibition and cited roadhouses where alcohol was served openly.[45]

The article identified Deputy Sheriff William McGuirk as a prominent member of the Knights of Columbus. The Knights were an organization founded in New Haven, Connecticut, to help immigrants and to fight discrimination. This meant that, like the Hulmans, McGuirk was Catholic and an immigrant.[46] The Knights published anti-Klan books that supported immigrants and African Americans. The American Unity League publication, *Tolerance*, was also anti-Klan. They broke into Klan offices, stole membership lists, and printed the names of Klan members—"the nightie-clad koo-koos"—under the heading "Who's Who in Nightgowns."[47]

The War with the Chain Stores

The Hulmans faced another threat. As Americans began to push into the countryside and expand city perimeters, America's food-buying habits changed. Local family grocery businesses like Hulman & Co. were in decline. Chain stores like A&P were able to take advantage of economies of scale in a way that companies like Hulman & Co. could not.

A&P began in 1859 as the Great American Tea Company. It was not a brick-and-mortar grocery, but a mail-order tea and spice business. The strategy was profit through volume: keep prices low and sell in volume. In 1870 the company capitalized on the completion of the transcontinental railroad the previous year and rebranded itself as the Great Atlantic & Pacific Tea Company—A&P for short.[48] In 1885 A&P came out with its first private-label product: baking powder.[49] Expansion of the product line and storefronts created massive, explosive growth. Small, independent stores could not compete and suffered losses.

By the 1920s the National Association of Retail Grocers and local organizations across the United States were fighting the "chain store menace," which controlled 40 percent of the grocery trade.[50] Indiana tried to stem the tide by taxing chain stores, but "the chain stores' share of market in many lines increased more quickly than elsewhere despite the taxes."[51] By 1930 A&P was doing more than one billion dollars in business and had 15,418 stores, more "than any other chain-store company in any product line has done before or since."[52] In 1919 retail margins were lower in chain stores (18 percent) than in independent groceries (19.5 percent), but both were eclipsed by the highest-margin stores: jewelry, 38.2 percent; household appliances, 37 percent; cameras, luggage, toys, and sporting goods, 35.2 percent. Liquor and candy were tied for fourth place, both at 35 percent.[53]

In 1926 Hulman & Co. wrote and distributed a flyer called "Foreign-Owned as against Home Owned Stores." Although it was ostensibly aimed at the chain stores, the flyer also countered Klan anti-immigrant propaganda. It positioned Hulman and other companies owned by immigrants or the descendants of immigrants not as foreigners, but as neighbors and good Americans. The new advertising campaign stressed that home-owned stores, unlike chain stores, put money back into the community.[54] They paid taxes in Terre Haute, patronized banks in Terre Haute, supported charities in Terre Haute.

In 1927 Hulman & Co. became affiliated with the Chicago-based Independent Grocers' Alliance of America, a nationwide organization of wholesale grocers who shared information to combat the encroaching chain stores. The Hulmans investigated other Alliance stores and on September 17, 1928, informed their sales force of a new plan:

> Times change—so do business methods. The old way was to keep a counter between your customers and your wares. The new way is to move your counters to the center of your store to give your customers access to the merchandise and serve themselves. . . . This naturally will demand price tags on each article and these must compare favorably with the prices of Chain Stores.[55]

No more open barrels of food with flies swirling around them. Instead, buyers roamed the aisles at will and helped themselves to hygienically packaged brand-name foods with individual price tags on open shelves. Also like the chain stores, Hulman & Co. ran weekly sales advertisements in newspapers and used window banners to advertise sales. These changes proved successful for Hulman & Co. for a while.[56]

General Foods and Standard Brands

While Clabber Girl battled the KKK and A&P, Royal and Calumet were still battling each other and the FTC. After the FTC decision, both Royal and Calumet published books that claimed victory. In 1927 Royal produced *Alum in Baking Powder*, which contained the text of the Federal Trade Commission trial examiner's dismissal. Calumet countered with an appeal to the medical profession with "a booklet . . . attempting to prove that the potassium tartrate present in food prepared with cream of tartar baking powder will produce Bright's disease," a kidney disease.[57] The following year, Calumet published *The Truth about Baking Powder*, a rehash of the FTC report. It included the original charges against Royal and the FTC's reinstatement of the charges. The baking powder war showed no signs of abating.

In 1927, Ziegler made a $1 million gift to the Harvard Business School to endow the William Ziegler Chair in Industrial Relations.[58] This was Harvard's second endowed chair directly connected to baking powder, after the Rumford chair in the chemistry department, which Eben Horsford had held. But Ziegler did not leave Harvard empty-handed; he hired the assistant dean of the school as executive vice president of Royal.[59]

There was a great deal at stake in the baking powder war. In 1927 the United States produced 155,941,240 pounds of baking powder valued at $29,519,375. The major market for exports was British South Africa, which received 696,640 pounds. Other major markets, in descending order of importance, were Mexico, Great Britain, Argentina, the Philippines, and Sweden.[60]

Royal's and Calumet's share of that market shifted during the 1920s. Royal was in the lead at the beginning of the decade. Calumet started to surge ahead in 1924, when it began its "double acting" advertising campaign. Royal made a comeback in 1925 and continued to hold its own in 1926, when the two companies were neck and neck. However, in 1927 Calumet surged ahead of Royal and leaped even more in 1928.

In 1928 the baking powder war ended abruptly. It was not ended by the USDA, the U.S. Congress, the president, the Remsen Board, the FTC, the U.S. judicial system, the national press, female consumers, or any other entity that had been involved in the war for more than half a century. It ended because J. P. Morgan & Co. decided to enter the food business, just as it had organized the railroads in the nineteenth century and United States Steel, General Electric, and AT&T at the beginning of the twentieth. Warren Wright sold Calumet to the Postum Cereal Company. Selling Calumet was not the elder Wright's idea. In fact, he was dead set against it. His son negotiated the deal behind his back, and had to work hard to convince his father to let the company he had created go. For thirty-two million dollars, Wright agreed.[61]

Postum was a new type of company, a conglomerate. These were large, diversified firms that combined companies that made different products but in the same

Table 9-1. Calumet Overtakes Royal

	Gross		Net	
Year	Calumet	Royal	Calumet	Royal
1923	5,660,686	4,971,118	1,622,848	2,189,188
1924	5,913,281	4,733,585	1,392,822	1,562,333
1925	5,727,193	4,663,533	1,540,272	1,359,201
1926	5,722,673	4,775,449	1,874,499	1,885,282
1927	5,795,386	4,931,459	2,022,049	1,487,431
1928*	3,356,583	2,435,093	1,212,971	725,769

*Six months ended June 30.
Source: "How Baking Powder Earnings Compared," *Wall Street Journal,* Jan. 5, 1929, 4.

type of business. C. W. Post, Kellogg's chief competitor, had founded Postum in 1895 to manufacture the cereals he created, especially Post Toasties and Grape Nuts. Postum had also acquired Minute Tapioca, Maxwell House coffee, and a seafood company. In 1929 the Postum Company acquired a new name: General Foods. Chase National Bank, a Morgan ally, was one of the forces behind General Foods, which had assets of seventy million dollars.[62]

Also in 1929, J. P. Morgan & Co. created another food conglomerate. Standard Brands combined Royal Baking Powder, Fleischmann's Yeast, and Chase & Sanborn coffee. Standard Brands had assets of $67,500,000 and "a stock market value in excess of $430,000,000."[63]

On June 12, 1929, the FTC issued its last proclamation regarding the Royal–Calumet baking powder controversy. Order No. 1127 was a cease and desist order to Calumet to stop the water glass test that its sales people and demonstrators had been using for decades. The commission found that "the extent to which the foam mixtures rise in the cold water glass test is not indicative of the comparative leavening strength of powders." They also agreed that, Calumet's protests to the contrary, the test was not used by grocers and housewives to determine if baking powder had deteriorated. Press favorable to Royal gloated: "'Trick' Selling Condemned by Federal Trade Commission."[64]

General Foods and Standard Brands radically altered both the way baking powder was advertised and the function of the individual companies they acquired. Both General Foods and Standard Brands subsumed individual companies into the conglomerate. No longer were Royal and Calumet pitted directly against each other. No longer were baking powder formulas an issue. The conglomerates ended the advertising strategy that William Ziegler had begun in the 1870s and that had proven successful for more than half a century. Nineteenth-century Wild West shootouts between individual gunslingers were not a productive way to wage war when conglomerates were building modern twentieth-century armies based on teamwork. Instead, both baking powders were cross-promoted with and tied to other food products within their companies. This increased advertising exponentially, because whenever a consumer saw an advertisement for one product, it was often connected with the others. Tying companies together created a food "family" and the idea that these foods went particularly well together. It also reinforced the identity of the conglomerate as a brand in and of itself. General Foods informed the public of its acquisition of Calumet in ads that said, "The Makers of Baker's Breakfast Cocoa, Minute Tapioca, Swans Down Cake Flour, Maxwell House Coffee, and Jell-O now bring you Calumet, the double-acting baking powder." It was a distinctly American business model: *e pluribus unum*—out of many, one. The virulent, decades-long advertising war between the cream of tartar and alum baking powders was over.

The end of the advertising war; General Foods introduces Calumet. Kraft Heinz Company.

At the beginning of 1929, there were high expectations for the economy. Herbert Hoover's inauguration held the promise of continued prosperity that the Republican National Committee had made on his behalf during the campaign: a chicken in every pot.[65] Wall Street was bullish. However, at the end of 1929 the stock market crashed. During the economic depression that followed, the baking powder war, like the Coca-Cola–Pepsi war and the Ford–Chevrolet war, would be a cutthroat price war.

CHAPTER 10

The Price War

The Fight for the National Market, 1930–1950

A good article will take care of [selling] itself.
—Herman Hulman, Clabber, *House of Hulman*, 1900

Buy one, get one free.
—Anton Hulman Jr., Clabber Girl, 1931

The Depression restructured the baking powder industry. The baking powder economic growth bubble, like the seventeenth-century tulip bubble in the Netherlands and the eighteen-century South Sea bubble, burst. No longer was baking powder production automatically profitable. The number of baking powder manufacturers plummeted as the Depression caused a scramble for market share and a price war. The competition for those consumers who had the limited amount of money in circulation was fierce. Businesses had to learn new ways to market existing products or to create appealing new products, or they went out of business. The New Deal brought government involvement and unions into the baking powder industry. During World War II, as millions of women streamed into the workforce, homemade bread production plummeted while commercial bread production rose. The result of these changes was that what had been a two-horse race between Royal and Calumet in 1930 became an open field with a dark horse in the winner's circle in 1950.

American business ingenuity went into overdrive during the Depression as Americans continued to consume sweet baked goods. Food inventions and innovations included Pepperidge Farm breads, Bama pies, and Toll House cookies, all the creations of women. Marge Rudkin named her above-market-priced specialty breads Pepperidge Farm after her Connecticut home. Cornelia Marshall realized that although people might not be able to afford entire large pies, they could afford small, personal-size pies, and named her company Bama Pies after her Alabama roots. Ruth Wakefield supposedly invented Toll House cookies by accident in Massachusetts in 1937; two years later Nestlé marketed the chocolate morsels, and the chocolate chip cookie was on its way to becoming an American icon.[1] In 1934 the Girl Scouts began their fund-raising cookie sales.

In the baking powder industry, changes and innovations included a new cookbook, a new biscuit mix, a new advertising slogan, new leadership, and a new war. All of them began in 1931, a watershed year for the industry.

The Joy of Cooking

The new cookbook was *The Joy of Cooking*. In 1931 another woman on the margins revived the tradition of the personal cookbook. This was an act of desperation by a fifty-three-year-old German American widow in St. Louis who found herself in what was politely referred to then as "reduced circumstances"—staring economic disaster in the face—after her husband's suicide. Irma Rombauer used the last of her savings to self-publish three thousand copies of a small book, *The Joy of Cooking*. Its subtitle, *A Compilation of Reliable Recipes with a Casual Culinary Chat*, perhaps subconsciously echoed the Reliable Recipe series of cookbooks issued by Calumet. It also brought the female voice and advice based on home kitchen experience back into cookbooks. Rombauer's daughter, Marion, a Vassar graduate, illustrated the book. In 1936 Indianapolis-based publisher Bobbs-Merrill took *The Joy of Cooking* nationwide. Its only competition was two cookbooks originally published in the nineteenth century, *The Boston Cooking School Cookbook* and *The White House Cook Book*, "neither of which had been updated in years."[2] Since then, through eight editions, *Joy* has sold millions of copies.[3]

The Joy of Cooking reveals how widespread baking powder was by 1931. More than a quarter of the recipes in the book contain baking powder, 95 pages out of 377.[4] Cakes made with yeast, the default a hundred years earlier, were now the exception, so unusual that they featured the word "yeast" in the title. However, *The Joy of Cooking* also reveals the problems that nonstandardized baking powder still caused. Rombauer does not mention brand names, simply generic "baking powder." From the proportions, it is clear that the baking powder Rombauer used was cream of tartar–based. She neglected to mention this, and it caused her much grief in subsequent editions. Using a sodium aluminum sulfate baking powder in the amount intended for a cream of tartar baking powder produces a misshapen, inedible product. Muffins will rise too much, peak lopsidedly, and taste bitter.

In the 1936 edition, Rombauer attempted to rectify this oversight with descriptions of the various brands and stepped into the baking powder war crossfire. On September 14, Rumford sent a letter of protest to *Joy*'s publisher claiming that

Rombauer's description of Rumford was "unfair." Rombauer sent a letter of apology, stating that she was "an amateur writing for amateurs."[5] Beginning with the second printing in 1936 and continuing into the 1950s, Rombauer added a two-page introduction to the baking section that tried to make sense of the various baking powders and to mollify Rumford in her new description of calcium phosphate baking powder:

> When confronted with the questions growing out of the use of the various forms of baking powder now on the market the puzzled layman is apt to sigh for the good old days when this article was made at home, rather haphazardly, from a formula handed from one generation to the other.
>
> . . . Due to the complexity of the problem it became one of life's major issues.
>
> . . . I had to battle with many new things, to discover that calcium phosphate baking powder while not necessarily so labeled has a double action, that both calcium and tartrate baking powders may be used in smaller amounts than usually designated in recipes without other harm to the baked product than a slightly smaller volume, that when eggs are added to a batter the usual measurement of one teaspoonful of baking powder to a cupful of flour may be reduced, etc., etc.
>
> I do not pretend to have solved the baking-powder problem scientifically but endless experiments have enabled me to solve it to my own present satisfaction.
>
> . . . Look at the label of your baking-powder can. There you will find distinctly marked the type of baking powder you are using.
>
> The recipes in this book will tell you plainly how much of each type of baking powder I find it advisable to use to make a successful bread or cake.[6]

Then, in each recipe that called for baking powder, Rombauer indicated how much of each type to use, one below the other. It immediately became apparent that alum baking powders, which she called "combination-type," were more powerful than tartrate baking powders like Royal. For example, the basic biscuit recipes call for two teaspoons of "combination-type" but four teaspoons of tartrate baking powder.[7] Throughout World War II and into the 1950s, while *The Joy of Cooking* used this format, it inadvertently promoted alum baking powders.

The Bisquick Manifesto

Also in 1931, General Mills introduced Bisquick, a revolutionary new biscuit mix with a phosphate baking powder built in. Bisquick went beyond self-rising flour because it also contained powdered milk, like pancake mixes. But Bisquick went beyond pancake mixes, too, because it contained a crucial ingredient that was new to mixes: shortening. General Mills experimented for months to find a way to incorporate a fat that would be shelf-stable and not go rancid quickly. The secret "Ingredient S" they hit upon was sesame oil. The other secret was the packaging.

Packaging the mix in cardboard boxes like other mixes did not work; paper was porous and allowed exposure to air, which oxidized fats and made them go rancid. Since fats were the problem ingredient, General Mills packaged Bisquick as if it were butter, in nonporous parchment.[8]

This single innovation of added fat got rid of the bane of biscuit making: "cutting in." To make good biscuits, the shortening has to be cut in to the dry ingredients until it is the consistency of coarse meal or the size of small peas. This has to be done by hand, with a snapping motion of the fingertips, or with two knives, or with a wire contraption called a pastry blender. It is time-consuming, and in 1931 there was no kitchen appliance, hand-operated or electric, that could perform this function. A food processor can do this now, but purists still believe in the finger-snapping method.

Cutting in is also where the variables that lead to failure lie. If the hands are too warm, the shortening will melt. If the shortening is not cut down to the right size, if the weather is too hot, if the dough is handled too much—all of these can produce soggy, tough, or leaden biscuits. But with Bisquick there was no cutting in, no failure. The front of the box instructed, "Add water or milk—nothing else," and delivered what it promised: "90 SECONDS from package to oven."[9] Bisquick, with its "Miracle Way of Making Biscuits," was an instant sensation. It sold more than half a million cases in its first seven months and spawned scores of imitators.[10]

General Mills used America's Sweetheart, Hollywood superstar Mary Pickford, as its spokesperson. The golden-haired Pickford had starred in upbeat silent movies like *Pollyanna* (1920), *Rebecca of Sunnybrook Farm* (1917), and *The Little Princess* (1917). The first recipe in the 1933 cooking pamphlet *Betty Crocker's 101 Delicious Bisquick Creations* was for strawberry shortcake, supposedly Pickford's favorite:

STRAWBERRY SHORTCAKE

2 cups Bisquick
2 tablespoons sugar
¾ cup cream

Bake at 450F for 12 minutes.[11]

Bisquick also claimed to be the cure for "Bride's Biscuits," those feeble attempts by the newlywed housewife to make biscuits like the ones her husband's mother made. Bisquick promised that its biscuits—or as it called them, "Bisquicks"—would make men happy.

Betty Crocker's 101 Delicious Bisquick Creations presented Bisquick as a marvel of scientific efficiency. This was a new field of engineering, also known as "Taylorism," after Frederick "Speedy" Taylor, its founder. Other pioneers in motion study and scientific management were Frank and Lillian Gilbreth, the couple who used their own efficiency principles at home with their twelve children, made famous in the book and 1950 movie *Cheaper by the Dozen*. Bisquick advertising applied industrial measurements of efficiency and precision to home cooking: Bisquick was "as much as 108% faster." For pancake making, "the new-fashioned BISQUICK method was exactly 83.75% faster than old-fashioned methods, and . . . it saved 45.5 steps."[12] Bisquick also saved time because, unlike flour, which had to be sifted, the instructions for Bisquick, printed in bold, were a pleasure to every housewife's eyes: **Do not sift Bisquick.**[13]

Bisquick not only saved time; it was also inexpensive. It cost "less than 1/7c (one-seventh of a cent) more than the old methods." This included fuel and the housewife's labor, figured at twenty-five cents an hour, "although we think it is too little." All of these figures were qualified by a "Certified Public Accountant . . . as Representing Results Obtained in Average Kitchen." Bisquick was marketed as new technology. Like "the vacuum cleaner, the electric washing machine, [and] the electric iron," Bisquick promised to free women from the "slavery of housework."[14]

Perhaps because not all women viewed homemaking as slavery, General Mills removed the "slavery of housework" from its 1935 cooking pamphlet, *How to Take a Trick a Day with Bisquick*. Subtitled "As told to Betty Crocker by screen stars, society stars, home stars and star homemaking editors," it capitalized on the new mania for playing bridge and featured pictures and recipes from Hollywood stars Joan Crawford, Bette Davis, Bing Crosby, Dick Powell, and others. This was pan-Hollywood participation, because the stars were from different studios. The "home stars" included foreign princesses along with social luminaries from Kansas City and Grosse Point Farms, Michigan, every one ostensibly a Bisquick devotee.

Bisquick saved time and was inexpensive, but that did not mean it made only cheap food. According to its cooking pamphlets, Bisquick could be used for inexpensive dishes like "Pigs in a Blanket" (small hot dogs rolled in a biscuit), corn bread, or chicken pie. But Bisquick could just as easily be used for creations suitable for a "Bridge Luncheon," a "Hunt Club Breakfast," a "Yachting Luncheon," or a "Golf Club Buffet Luncheon." Bisquick was also a favorite with manly men like movie star Clark Gable, who ate Bisquick griddle cakes with maple syrup, or outdoorsmen like Dr. A. E. DuBarry of Lyons, Montana, who prepared Bisquicks "in the camp stove, over hot coals, in a Dutch oven, and between two frying pans."[15]

What neither Bisquick cookbook had was images of African Americans, except as servants. This was ironic because the idea for Bisquick had come from an African American. According to the General Mills official history, in November 1930, Carl Smith, an executive of Sperry Flour, a General Mills West Coast affiliate, was on a Southern Pacific train. Lunchtime was long gone, so Smith was amazed when fresh hot biscuits appeared on the table in what seemed an impossibly short amount of time. Smith went into the kitchen, where the African American male chef explained that he had prepared everything for the biscuits—flour, baking powder, salt, shortening—in advance, so that all he had to do was add liquid. This

DICK POWELL

Warner Brothers Star of
"SHIPMATES FOREVER"
AND OTHER SUCCESSES

● is proud of his bachelor establishment and loves to fix

Impromptu Suppers

like this one—

Chicken a la King (Canned)
in Bisquick Ring
Celery Olives Pickles
Fresh Fruit Butterscotch Cookies
Coffee

or this—

Mexican Rarebit over
Split Bisquicks
Coffee Tossed Salad
Devils Food Cake

See Bisquick recipes on Pages 36 to 41

★

● The kitchen in Dick Powell's new home is as practical as it is attractive—decorated in a cheerful yellow and white, with windows on both sides. He himself designed the movable work table with heavy top, and cupboard and drawer space underneath to hold supplies. He proudly shows it to all visitors and sometimes takes his guests into the kitchen to eat. On Cook's night out, he wades in and tries his hand at making his pet dishes. Dick brags about his cooking.

Dick Powell's original idea to spoon Bisquick Dough into a greased ring mold.

Actor Dick Powell gets creative with Bisquick in the kitchen. *How to Take a Trick a Day with Bisquick,* General Mills, Inc., 1935. Author's collection.

practice of every good chef is called *mise-en-place,* French for having everything ready before you begin to cook. Smith took the idea of a prepared biscuit mix back to General Mills, which refined it. The name of the conscientious and innovative black chef is not known.[16]

African Americans were not pictured at all in the 1933 *Betty Crocker's 101 Delicious Bisquick Creations* but were included twice in the 1935 *How to Take a Trick a Day with Bisquick,* both times as servants. A photograph showed a black male in a white shirt and black bowtie as he served food to actor William Gargan and his wife.[17] The other reference is a one-and-a-half-line quote from a fictitious Aunt Jemima–type servant—"Mammy Lou says: 'Laws, Honey, it don't take no time at all. I use Bisquick batter to fry my chicken in, and folks tell me my chicken has the real old Southern flavor.'"[18] Mammy Lou was in a two-page spread aimed at the Southern market, which the earlier Bisquick cooking pamphlet had overlooked.

If Bisquick took the baking mix market by storm, Jiffy took it by stealth. According to the official Jiffy history, in 1930 a Michigan woman named Mabel Holmes had an epiphany when one of her twin sons brought another boy to the house. The boy was being raised by his father, who had packed a hard, dry, flat biscuit for him to eat—completely unpalatable. Mabel decided that she could fill the need for a way to make biscuits that was "so simple that even a man could do it."[19] In 1930 Jiffy Baking Mix was carried by one grocery chain in Michigan. It expanded from there.

Unlike Bisquick's highly visible, expensive advertising campaign, which included glossy cooking pamphlets and movie stars, Jiffy took a unique approach to advertising: it didn't. This cut Jiffy's overhead costs, savings it passed on to the consumer. Jiffy still spends nothing on advertising. Since advertising is generally 33–35 percent of a food item's cost, even large supermarket chains like the modern Kroger cannot compete with Jiffy's price.

Calumet, the Double-Acting Baking Powder

In 1931 General Foods overhauled Calumet's advertising. During the 1920s, Calumet had a picture of a Native American chief in full feathered headdress on the can, but it used a Kewpie doll in its print advertising and on its cookbooks. It had "Not Made by a Trust" on the label on the can. General Foods retired the Kewpie doll. Then it removed the words "Not Made by a Trust" from the label on the can. Calumet continued to use the water glass test but modified it. The company no longer claimed that the test was proof of Calumet's superiority to other baking powders. It was just proof of Calumet's double action.

TEST CALUMET'S DOUBLE-ACTION THIS WAY

Put two level teaspoons of Calumet into a glass, add two teaspoons of water, stir rapidly five times and remove the spoon. The tiny, fine bubbles will rise slowly, half filling the glass. This is Calumet's first action—the action that takes place in the mixing bowl when you add liquid to your dry ingredients.

After the mixture has entirely stopped rising, stand the glass in a pan of *hot* water on the stove. In a moment a second rising will start and continue until the mixture reaches the top of the glass. This is Calumet's *second* action—the action that takes place in the heat of your oven.

Make this test. See Calumet's *Double-Action* which protects your baking from failure.[20]

General Foods also began advertising that capitalized on Calumet's chemical reaction. In 1931 the words "Double-Acting" appeared for the first time prominently on the front of the can, above "Calumet," where "Not Made by a Trust" had been.

The Baking Powder Price War for the National Market

The year 1931 was a watershed year for Clabber Girl too. Leadership of the company passed to the third generation of the Hulman family. Anton Hulman Sr. stepped down and gave the title to his son, Anton Hulman Jr.[21] Tony Hulman, born in 1901, had graduated from Yale's Sheffield Scientific School with a degree in engineering. He was also a world-class athlete and drawn to speed. He was in two motorcycle accidents before he was in his teens, and was "best school-boy [pole] vaulter in the U.S." in 1919. He was voted into the American Athletic Union (AAU) while still in East Coast prep high schools at Lawrenceville, New Jersey, and Worcester, Massachusetts. While Tony played basketball and baseball, he really excelled at track and field: hurdles, high jump, long (broad) jump, shot put. He continued in track and field at Yale—ten medals in pole vaulting and high hurdles in his freshman year alone, and then All-American in those sports. Tony still found time to row varsity crew and, most importantly, he was offensive end on the Bulldogs' legendary undefeated 1923 and 1924 football teams. He lettered in seven sports and was voted best all-around athlete at Yale.[22]

Nevertheless, Anton Sr. had insisted that Tony learn the business from the ground up, so Tony was one of Clabber Girl's first drivers when the company switched from horses to motorized trucks in the 1920s. Over the following years, Tony worked in every department of the company. He was a "regular guy," who got along with his coworkers.[23]

On October 6, 1926, Tony got married and then embarked on a four-month European honeymoon. He and his wife, Mary, came back to live at 1327 South Sixth Street, and on November 4, 1930, their daughter Mary was born. Unfortunately, the infant lived only a few hours.[24]

In the fall of 1931, on the heels of his recent personal tragedy, what changes did Tony Hulman see that would benefit Clabber Girl? What entrepreneurial vision did he have when he took stock of his company and its prospects? The United States was two years into what was clearly just the beginning of a long economic downturn. Unemployment was climbing, people were losing their homes, farmers losing their farms. A new type of store was added to the threat from chain stores. Just as stores like A&P threatened independent groceries, the new supermarkets began to threaten A&P and other chain stores. The supermarkets, huge stores, mushroomed in the suburban landscape, where land was inexpensive. This kept overhead low and made free parking available as American consumers traded in horses and public transportation for the privacy and convenience of automobiles.

Supermarkets were a radical departure in food merchandising. They "had even higher volume, higher turnover, and lower margin than A&P," whose stores were located in older cities.[25] Inside the supermarket there was no grocer or clerk between purchaser and product. No longer did customers have to line up at a counter to tell the grocer what they wanted and then wait while he went and got it for them.

Supermarkets democratized food purchasing by cutting out the middle man: the grocer. Removing the grocer had tremendous implications for food manufacturers. Soon they would not be able to rely on grocers to sell goods for them. Instead, consumers would buy national brands, names they recognized from advertising in the national press and the radio, and which used economies of scale to sell at lower prices. In a few years these convenient new supermarkets would take over the country. No doubt Tony Hulman understood the implications of this: the business his grandfather had founded eighty years earlier was at stake. So was Clabber Girl. If Clabber Girl did not become a national brand, and soon, even its local market would be gone. Clabber Girl would disappear, squeezed out by national brands.

"Entrepreneurial drive and vision have been essential to creating and organizing mass markets."[26] This, Tony Hulman had. What was Clabber Girl facing? What was Hulman's marketing strategy? Perhaps he was encouraged by the Continental Baking Company in Indianapolis. Continental was having nationwide success with two new products: a small, cream-filled sponge cake called Twinkies, and Wonder Bread, which it had begun selling sliced, a wildly popular innovation, at the end of the 1920s.[27] Tony Hulman had confidence in his baking powder, a product that had proven itself with consumers for more than thirty years.

To enter the national market, Clabber Girl would have to go head-to-head with firmly entrenched companies with established consumer recognition. Rumford had been in the baking powder business as long as Hulman & Co. had been in existence. Calumet, also an alum baking powder, had been the market leader for decades. Royal was the darling of home economists. All three companies had been publishing cookbooks for decades, Royal since the 1870s. Rumford, Calumet, and Royal all had established distribution networks, regional offices, agents, and media contacts throughout the United States. Long-standing relationships with grocers meant that they could command prime shelf and display space. Royal and Calumet were part of conglomerates with extremely deep pockets and resources. Any losses they suffered could be offset by the other companies in the conglomerate. Against them, Clabber Girl had six trucks and a local sales force.[28]

Clabber Girl also had to compete with the lowest-priced baking powder, KC, produced by Chicago-based Jaques Manufacturing. The price of KC had held steady at one cent per ounce for decades; their advertising slogan was "25 ounces for 25 cents." Clabber Girl met KC's price but decreased the size of the can, because many people did not have a quarter to spend on a twenty-five-ounce can. Clabber Girl began to advertise in national publications, "10 ounces for 10 cents"—the same price as a cup of coffee in the famous Depression song "Brother, Can You Spare a Dime?"

Then Clabber Girl beat KC's price. Tony Hulman decided to use a marketing strategy that had proven successful for Clabber Girl in Indiana, Ohio, and Illinois: he gave Clabber Girl away. But Tony Hulman wasn't giving Clabber Girl away to consumers, and he wasn't giving it away by the can. He was giving it away to grocers—by the *case*. For every case of Clabber Girl a grocer bought, he received a case free—pure profit at whatever price the grocer chose to sell. This made grocers across the country Clabber Girl's de facto agents at point of sale. It created an instant national sales force without adding sales personnel. Consumers now are used to "BOGO"—Buy One, Get One—but eighty years ago it was a new strategy.

In addition to a competitive price and giveaways, Tony Hulman made sure that Clabber Girl had something that only two other baking powders, Royal and Calumet, had: the *Good Housekeeping* Seal of Approval. This was a guarantee by the magazine, not the manufacturer. *Good Housekeeping* promised that if a consumer purchased one of its "Guaranteed Products" and complained to the magazine about it, the magazine would investigate. If the product was found to be defective, *Good Housekeeping* would either refund the purchase price or replace the product.[29]

Wiley had championed Royal at *Good Housekeeping* from the beginning. To give the Seal of Approval to a baking powder containing sodium aluminum sulfate was a major shift for the magazine. It would have been impossible while Wiley was there, but Wiley had died on June 30, 1930, and had left the magazine a few months earlier. Calumet wasted no time procuring the seal after Wiley's departure. Calumet Baking Powder appeared on the "Guaranteed Products" page for the first time in the January 1930 issue of *Good Housekeeping*. Clabber Girl was on the list by 1931.[30]

Clabber Girl's advertising was designed to challenge its competitors. Like them, Clabber Girl claimed that it was "a scientifically compounded private formula." Like Calumet, on the can, immediately below "CLABBER GIRL," was "Double Acting." Like Royal, Clabber Girl advertised, "Leaves No Bitter Taste." Like Royal, Rumford, and Calumet, Clabber Girl also began to publish cookbooks. In 1931 the cover of its first cookbooklet, *Clabber Girl: The Healthy Baking Powder*, was a picture of a rosy-cheeked girl with a double chin. This was a rare misstep on the part of Clabber Girl. Heft might have worked in the German market, but it was not attractive to weight-conscious Americans, even during the lean times of the Depression. The chubby child was retired and replaced by a thin, single-chinned woman. In 1934 Clabber Girl published a short promotional cookbooklet titled *Clabber Girl Baking Book*. The cover featured a smiling, svelte housewife.[31]

Clabber Girl's aggressive marketing had stunning success. On February 8, 1933, Rumford's annual shareholder meeting took place. The directors tried to be upbeat by reporting on the competition facetiously: "Clabber Girl has been the most active 'hussy' of them all, and in order to gain distribution and favor stepped out of her natural field of operations and sought other fields for conquest. Clabber Girl distributed her favors freely; giving her goods away where sales could not be made—actually giving a case free with each case purchased."[32] Rumford management's derisive personification of Clabber Girl as a whore masked real fear. Clabber Girl was making tremendous inroads all over the United States. It was clear that women were voting with their pocketbooks and buying the lower-priced alum baking powders in smaller containers. Rumford tried to compete by rushing a new brand, Health Club, its own lower-priced sodium aluminum sulfate baking powder, onto the market. Launching a new brand of baking powder into the Depression market, where brands that had been in existence for decades were struggling, was difficult at best.

In 1934, Jaques, too, came out with a new, lower-priced baking powder, called Gold Label. Jaques gave Gold Label away to jobbers, "shipping 50 to 100 cases without any charge whatever." In addition to giving the jobbers free baking powder, Jaques *paid* the jobbers 15 percent of Gold Label's list price. Jaques focused on the South, where Rumford was strong, and shipped Gold Label there by the carload.[33] Rumford felt that Jaques "would rather destroy the entire industry than allow anyone to make a legitimate dollar."[34]

Jaques' tactics were effective. Business dropped 73 percent in Rumford's Richmond, Virginia, office, so they closed it down. Sales were also down in the Baltimore and Chicago areas. Rumford closed their East St. Louis production facility, which had served the Midwest.[35] Another big blow to Rumford was the defection of its local customer base. New England sales were down 5 percent, but Rhode Island, Rumford's home, was down 20 percent, because it had been "very severely attacked by our competitor, Calumet."[36]

Rumford tried to reach consumers in new ways. It published cooking pamphlets, small all-purpose cookbooks that did not contain recipes for breadstuffs and desserts. Instead, they had recipes for vegetables and inexpensive Depression main dishes such as "Soybean Loaf," "Kidney Stew," and "Philippine Goulash." In 1929 Rumford produced *24 Uses for Rumford Phosphate Baking Powder in Cooking*. By 1932 the cookbook was in its third edition and had expanded to sixty-four recipes.[37]

Rumford also jumped on the Bisquick bandwagon and quickly brought out its own all-purpose mix, Bakes-All, a "combination of wheat flour, baking powder, dried egg, shortening, sugar and salt." However, Rumford was hampered by not being near a source of flour, so they contracted with a milling company in Alton, Illinois.[38] Unfortunately, Bakes-All showed a thirty-thousand-dollar loss in 1933.[39]

Rumford tried to cut costs in other ways. When railroad rates went up, the company switched to trucks for long-distance deliveries and managed to reduce freight costs slightly. They also instituted a 10 percent across-the-board pay cut on July 1, 1932.[40] That was not enough. In 1934 Rumford reduced its sales force from 134 to 90, which they estimated would save $150,000.[41]

All of the baking powder companies wooed grocers. KC gave grocers a free "Grocer's Want Book," in which they could keep track of their accounts.[42] Clabber Girl provided grocers with a chart showing how much baking powder they would have to sell, and at what price, in order to make a profit.

Rumford's subsequent reports to shareholders became less humorous. In the next four years, the company's business gained what the annual shareholder reports called "descent momentum" at a sickening pace. Production dropped from 17,861,387 pounds in 1930 to 9,325,197 pounds in 1934, a loss of 47.5 percent.[43]

There was also competition from a new type of baking powder. As commercial production of cakes and other baking powder–leavened baked goods increased, a new type of baking powder was necessary. Sodium acid pyrophosphate had a high bench tolerance—meaning it did not release most of its gas during mixing. Royal and the other cream of tartar baking powders did. Part of the decline in cream of tartar sales was because they were suitable only for household use. Machine production was a longer process because of the volume of batter being mixed, so it needed a slower-acting baking powder.[44]

Rumford looked to outside experts for help. They employed the Nielsen Marketing Service to get a different perspective. What Nielsen reported was a shock. By 1935, market share and power in the baking powder industry had completely shifted. Royal, which at its height had controlled most of the baking powder market, had less than 10 percent. Calumet had dropped to 19 percent from its one-third of the market in the 1920s.[45] The Depression and the price war had flipped the baking powder industry on its head.

As table 10-1 shows, Clabber Girl, KC, and Calumet—the top three alum baking powders—had 61 percent of the market. Royal is the only cream of tartar baking

Table 10-1. Baking Powder Market Shares, 1935

Percentage	Brand
23	Clabber Girl
19	KC
19	Calumet
9.2	Rumford
7	Health Club
7	Royal
5.5	Davis
10	Others

Source: RCW Annual Meeting, Feb. 12, 1936, 2.

powder, and the only baking powder in the trust, to make this list. Rumford's new Health Club brand did surprisingly well; it equaled the plummeting Royal in market share. Together, Rumford and Health Club had more than 16 percent of the market, which put them in fourth place, behind Clabber Girl, KC, and Calumet. But the price war was not over, and it was not the only factor that caused the baking powder industry shake-up during the Depression.

The New Deal

The baking powder war, the Depression, and the price war in the 1930s took their toll on the baking powder industry. By 1934, when the National Industrial Recovery Act (NIRA) codes for the baking powder industry were written, the number of baking powder manufacturers had dwindled from 534 in 1900 to 150.[46] Of those 150, only the 7 listed in table 10-1 had a market share of more than 5.5 percent. The other 143 companies shared the remaining 10 percent; each had a minuscule market share.

At the same time that its revenues were nosediving, Rumford's overhead was rising because of New Deal policies. President Franklin Delano Roosevelt's New Deal changed the relationship between business and government. Thirty-five years earlier, one baking powder company had bought a state legislature. Now, baking powder manufacturers, summoned not by Congress but by the president, came to the federal government as supplicants. In its report to shareholders in 1934, the Rumford board of directors shared their misgivings about the Roosevelt administration and its agenda: "possibly to redistribute wealth."

Under the New Deal's unprecedented NIRA, businesses were supposed to cease damaging competition and to draw up codes that set wages, hours, and good business practices. Because it was connected to food, even though it had nothing to do with farming, baking powder was originally under the aegis of the Agricultural Adjustment Administration (AAA). This did not go well. The baking powder manufacturers met among themselves and drew up a code, but rejected some of the changes the AAA requested, saying they were "repulsive to fair minded business men." This was because "unfortunately, A.A.A. has been dominated by three men of particularly radical views; . . . and it has been impossible to agree with them. They believe in many social changes and that business must get along without a profit." Protests to President Roosevelt resulted in the baking powder industry being shifted to the National Recovery Administration (NRA), which was supposedly more favorable to business.[47]

On April 30, 1934, leaders in the baking powder industry met with NRA representatives at the Ambassador Hotel in Washington, DC, to create a code. The chair was William Sweet, Rumford's treasurer. Although there were 150 baking powder companies in the United States, only 18 baking powder industry executives came to Washington. They controlled 65 percent of the baking powder market.[48] The men were hopeful that the government would force some of the destructive practices in the industry to stop—in other words, control competition. First was the price war and selling below cost. These were attacks on Clabber Girl and Jaques. The leaders also wanted to abolish gifts and rebates. Further, they wanted a prohibition on false advertising and disparaging remarks about the competition. This was directed at Jaques and Royal. Even though Royal had not engaged in these practices in several years, since it had become part of Standard Brands, the memory lingered and the fear remained. Some wanted to remove albumen from baking powder, because they believed it was there for only fraudulent purposes. This was an attack on Calumet. They also wanted an end to industrial espionage.

Also of great concern was the drop in home baking. These businessmen did not see the "one-third of a nation ill-housed, ill-clad, and ill-fed" that President Roosevelt talked about in his second inaugural address, in 1937. Instead, the upper-class, wealthy baking powder manufacturers attributed the decline in home baking to the new lives of leisure that women were leading. Attractions that drew women out of the home included "automobiles, motion pictures, swimming and tennis," women's clubs, the Parent Teacher Association (PTA), and especially the mania for bridge that was sweeping the country—"300,000,000 decks of cards were sold in 1927." There were distractions aplenty in the home, too: "the radio, the books, papers and magazines . . . compete for a share of their waking hours."[49]

Urbanization and easy access to commercial bakeries also took a toll: "Apparently the women of the cities have nearly all forgotten how to make bread (most of the younger women never knew)."[50] Cake, however, was another matter: "They still bake cake, of course." Approximately 80 percent of the sweet baked goods were still made at home. However, that was beginning to drop, too, as commercial bakeries constructed new factories that focused on cake.[51]

For the businessmen meeting at the Ambassador Hotel, women were the problem. Women's attitude toward and perception of their role differed from what the men thought it should be. "Their mothers—their grandmothers—accepted drudgery as the inevitable penalty of their sex. But these women have been educated against it."[52] The New American Woman could vote, drive, enter the professions. She was interested in convenience, and there was no turning back.

Housewives were a problem because they no longer wanted to bake. In the workplace, "girls" were a problem. How much to pay them? Let them work nights or not?

MR. WOODBURY (NRA): "It does not naturally follow that female employees get the same pay as male employees."

MR. SWEET: No."[53]

The baking powder companies, like other companies, found a simple solution to the "woman problem." They simply deleted females from the workforce wherever they could. No more messenger girls, only messenger boys. The code also made exceptions for workers based on age, physical handicaps, and mental handicaps. These workers "may be employed on light work at a wage below the minimum established by the Code."[54]

The tension between the baking powder manufacturers and the government representatives was barely controlled during the meeting. It was clear from the questions the government representatives asked that they did not understand the basics of the baking powder industry, or even of how manufacturing was generally done. When they tried to separate work hours for only the canning part of the business, manufacturers had to inform them that the hours had to be the same as the rest of the factory and to explain, in very simple terms, why. In continuous process manufacturing, the efficient American standard, interconnected machinery controlled manufacturing from start to finish. In baking powder factories, mixing the powder and filling the cans were synchronized, because the machines that mixed the baking powder poured it into the cans. The processes could not be separated without shutting down the factory.

The animosity erupted at the end. The baking powder manufacturers balked at allowing the government access to company records. The federal representative was taken aback, because this was standard in other codes, and "400 odd companies recognize this provision."[55] The other baking powder manufacturers backed up Chairman Sweet and dug in their heels. They said they did not see how the government could settle disputes that arose in the baking powder industry "if men who have spent their entire lives in the baking powder industry can't."[56] Baking powder executives were not the only ones who objected to the NIRA codes. A little more than a year after this meeting, in a rare unanimous decision, the Supreme Court declared the NIRA unconstitutional.

However, some of the constitutional provisions of the NIRA carried over into other New Deal legislation. Under the National Labor Relations Act, aka the Wagner Act, in 1935, employees had the right to collective bargaining. They could join a union if they wanted to, but were not forced to join. In the next ten years, union membership more than tripled, from 3.8 million in 1935 to 12.6 million in 1945.[57] Rumford workers were among them. On September 19, 1938, Rumford entered into a contract with the United Baking Powder Workers Local Industrial Union No. 456. The contract was for an eight-hour day, five days each week. There would be one week of vacation after one year. One year later, on September 19, 1939, the contract included the provision for a one-hour lunch, to be taken between 12:00 and 1:00p.m. "except in an emergency when the one hour period may come within the limits of 11:30A.M. to 1:00P.M."[58] Rumford's labor costs had started going up years earlier. In 1933 after Roosevelt's inauguration, Rumford's labor costs had already increased by twenty thousand dollars.[59]

The New Can: "Factory Fresh"

In 1934, the same year as the NIRA meetings, the Davis Baking Powder Company, based in Hoboken, New Jersey, brought new ammunition to the baking powder wars: a new baking powder can. The old slip-lid cans were wrapped in a paper label. To open the can, the label was slit. The top of the can rested on the bottom, without threads, so once the seal was broken, the can could not be resealed tightly. Davis's new metal can was airtight. The top could be easily popped off with the handle of a metal spoon and just as easily pressed back onto the can.

A second inner seal made of parchment added to the airtight seal. It was also something completely new, because it aided in measuring the baking powder. Some of the parchment was evenly sliced off so that a measuring spoon could be inserted into the can. When the spoon was withdrawn and scraped against the remaining parchment, it removed the excess baking powder neatly and provided

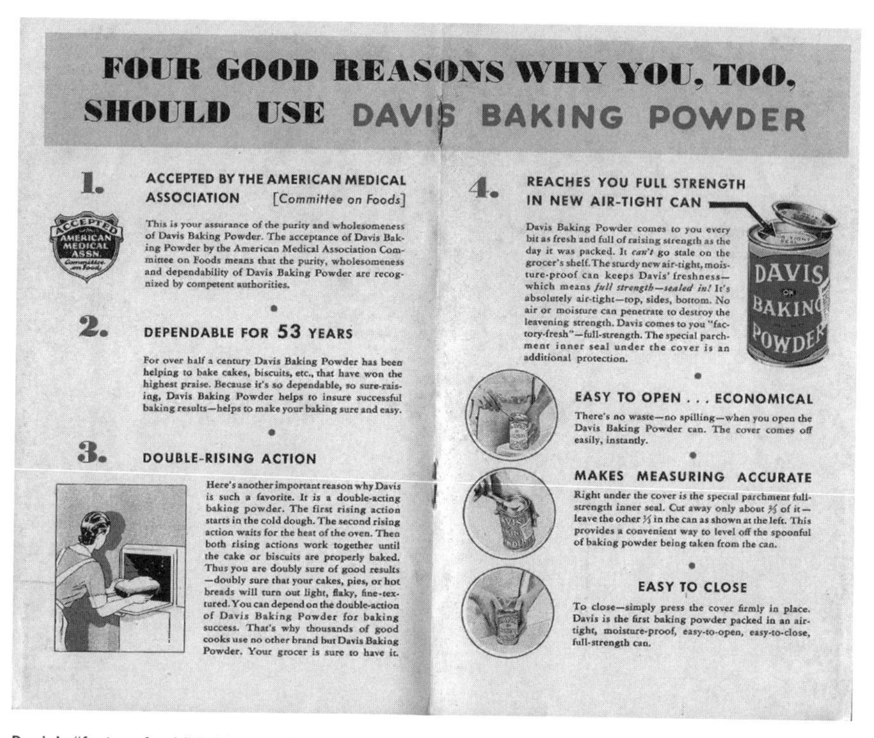

Davis's "factory-fresh" baking powder in a can with a new type of seal. R. B. Davis Company, Hoboken, NJ. Author's collection.

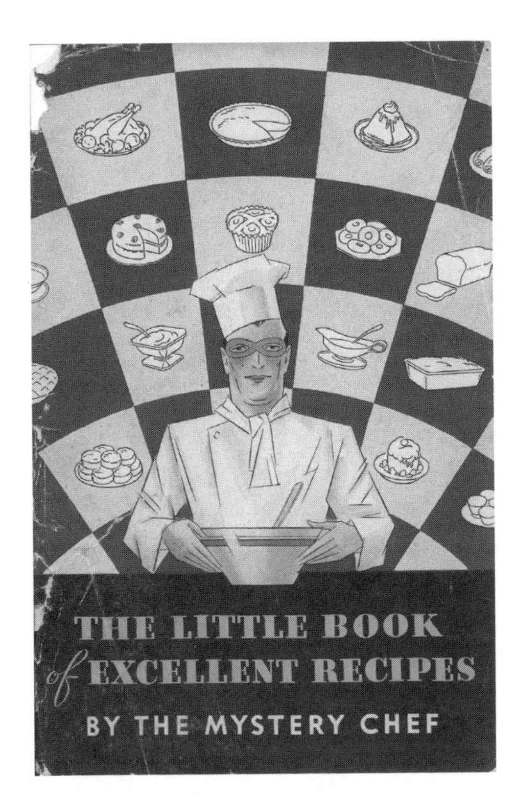

Davis's Mystery Chef, on the cover of its 1934 cookbook. *The Little Book of Excellent Recipes*, R. B. Davis Company, Hoboken, NJ. Author's Collection.

a level spoonful. Before this innovation, the spoon had to be leveled off with the back of a knife or spatula, which almost invariably resulted in spilled baking powder. Davis coined a phrase: the new can kept its baking powder "factory-fresh."[60] The can quickly became the industry standard. Calumet adopted the new can the following year.[61]

When Rumford's shareholders met in 1937, Jaques and Clabber Girl were still engaged in their cutthroat practices. Jaques was the sticking point: "Hulman & Co I am sure would be willing to cooperate quite freely if it were not for their fear of competition with the Jacques [*sic*] Mfg. Co."[62] Nine of the major baking powder manufacturers (minus Jaques) had consulted an outside agency, Gooch Jennings Co., Inc., to see what they could do to improve the market. The report results were a shock. Based on monthly sales and price reports from the nine companies, Gooch Jennings claimed that "the decline in dollar volume is not so much the result of decreasing consumption but is brought about by destructive practices within the industry itself." In fact, baking powder consumption had *increased* in the previous eight years.[63]

Rumford was also aware that part of its problem was that it was becoming obsolete in the business world because it was "a one-product company." General

Table 10-2. Baking Powder Consumption, 1927–1935

	Pounds Sold	Price
1927	102,331,183	$21,882,175
1935	104,914,779	$14,164,054

Foods, on the other hand, "acknowledge freely that Calumet Baking Powder has suffered greatly in sales but many of their other products are selling well, so that their total revenue from all products has not suffered." Rumford did not have the luxury, as Calumet did, to obtain nine hundred thousand dollars from a parent company to increase sales.[64] This would make the difference in which companies would weather the Depression and World War II.

Clabber Girl, too, bankrolled by the Hulman grocery business and diversification into real estate, among other areas, used some of the profits from the new national business to buy Forest Park, 360 acres for exclusive use as a park by Hulman & Co. employees. It had a clubhouse with a fireplace, comfortable chairs, magazines, parlor games, a radio, and a piano; a playground with swings and see-saws; picnic tables and outdoor cooking areas; row boats along a creek; and an artificial lake and swimming pool. The inaugural event, attended by six hundred employees, was a hunt for fifteen hundred Easter eggs on Easter Sunday, April 13, 1941.[65]

Anton Hulman Sr. had initiated the purchase of Forest Park. In December 1941 he left to spend the winter in Miami, Florida. His sudden death there on February 9, 1942, was a shock to everyone. He was buried in Calvary Cemetery just outside Terre Haute, along with his father, mother, sister, and brother.[66]

Cookbooks

The end of the 1930s saw a renewed interest in Southern cooking as cookbook writers and publishers rushed to sail on the whirlwind created by the 1936 blockbuster novel and 1939 movie *Gone with the Wind*. These cookbooks glorified hospitality in the antebellum South and therefore, to greater or lesser degrees, slavery. The cookbooks included Eva Brunson Purefoy's *Purefoy Hotel Cook Book*, which included recipes from the famous hotel in Talladega, Alabama; a modernized version of *The Martha Washington Cook Book*; an updated version of Mrs. Dull's 1928 *Southern Cooking; The Williamsburg Art of Cookery* from Colonial Williamsburg; and Wisconsin native Marjorie Kinnan Rawlings's *Cross Creek Cookery*, relating tales of her life in Florida, which followed her 1938 Pulitzer Prize–winning novel, *The Yearling*.[67]

In all of these books, the traditional baked goods that are supposed to be representative of the antebellum South are leavened with a product that had not existed then: sodium aluminum sulfate baking powder. Hot breads were central to Southern cuisine: "Except in cruelly troubled times [a euphemism for the Civil

War], no meal at Tara [the plantation in *Gone with the Wind*] (or any other Southern homestead) was complete without hot breads, served piping hot and dripping with butter."[68] *The Southern Cook Book of Fine Old Recipes*, published in 1939, contains racist stereotypical characters like "the old mammy, head tied with a red bandanna, a jovial, stoutish, wholesome personage." The bottom of almost every page has drawings of black people eating watermelon, fishing, stealing chickens, or lying on the grass with their shoes off. The drawings are accompanied by three- or four-line poems that call African Americans "darkey" and the "n" word. Many of the recipes contain baking powder: "Mammy's Baking Powder Biscuits," "Dinah's Rice Croquettes," "Confederate Coffee Cake," and "Pickaninny Doughnuts."[69]

The *Gone with the Wind Cookbook*, published in 1940, had a picture of the heroine, Scarlett O'Hara, on the cover, in front of Tara, the Georgia plantation where she lived. This cookbook is more interested in convincing the reader to have *Gone with the Wind* parties than in glorifying slavery.[70] Almost every recipe in the six-page section "Southern Breads" contains baking powder, including "Aunt Pittypat's Cream Scones" and "Atlanta Waffles."[71]

Rawlings's cookbook includes desserts that contain baking powder, baking soda, and cream of tartar. However, the cookbook is more notable for what she wrote about her maid, Idella Parker. To Rawlings, Idella was the ideal maid; to Idella, Rawlings was just another white racist who had to be obeyed. *Cross Creek Cookery* is drenched in racism. Rawlings compares African Americans to dogs, apes, and savages. Parker worked for Rawlings for fourteen years and did most, if not all, of the cooking when the recipes were tested for *Cross Creek Cookery*. She even provided many of the recipes but received credit for only three. One of these was "Idella's Biscuits."[72]

IDELLA'S BISCUITS

2 cups flour
4 teaspoons baking powder
¾ teaspoon salt
⅓ cup solid vegetable shortening
about ¾ cup of milk

1. Preheat oven to 450. Sift flour, baking powder, and salt together.
2. Using a fork mix again. Cut in Crisco with a fork.
3. Add milk and combine with your fork.
4. Roll onto pastry cloth or work area. Handle this dough as little as possible to make a tender, crisp biscuit treat. Roll out to ¼ inch thickness. Cut into very small rounds.
5. Bake in hot oven 12 to 15 minutes. Serve hot with butter.[73]

The South and the racism were present in other advertising. Calumet began to advertise in another particularly American form: black-and-white comic strips in men's and women's magazines. The strips took a full magazine page and contained six panels of two columns and three rows. Each self-contained story was a cautionary tale about how Calumet helped a housewife overcome her fear of baking because Calumet was foolproof. Like Bisquick, Calumet used a Hollywood tie-in. Singer Kate Smith, best known for belting out "God Bless America," was the Calumet spokesperson. A Kate Smith character sometimes appeared in

Overt racism increased in the 1930s because of the popularity of *Gone with the Wind. Country Gentleman*, c. 1939. Author's collection.

the Calumet comic strips to advise housewives. In one of them she showed that Calumet was so good that it put even a beloved Mammy's baking to shame.

World War II

World War II caused a seismic shift in women's relation to cooking, especially breadstuffs. Consumption of commercially baked breadstuffs increased dramatically. Between 1939, when commercial bakers produced about two-thirds of the total bread consumed in the United States, and 1945, when they produced between 80 and 85 percent, American women became accustomed to buying bread.[74] There were several reasons for this. Many housewives became factory and office workers during the war. The number of women employed outside the home increased 43 percent from 1941 to 1945. More people needed bread because they were eating lunches out of the home, often sandwiches, quick to prepare and with versatile, inexpensive fillings. World War II employment also led to an enormous increase in income: 61 percent from 1941 to 1945.

At the same time, home baking decreased because women had less time at home, and because bread ingredients like butter, sugar, eggs, and milk were rationed for home use after May 1942. Overall, it was more convenient for working women with a decrease in time and an increase in income to buy bread instead of baking it themselves. Between 1941 and 1945, baking industry output rose more than 50 percent, from 11,304,861 tons to 16,978,188 tons.[75]

However, even with wartime rationing, housewives still cooked other foods when they could. They looked to cookbooks for help, gobbling up wartime editions of Irma Rombauer's *The Joy of Cooking*—332,382 copies by 1946. During World War II, *Joy* eclipsed its main competitor, Fannie Farmer's *Boston Cooking School Cookbook*.[76]

In postwar America the percentage of bread produced by commercial bakeries continued to rise, as well as the percentage of sweet baked goods. By 1950, bread baking in the home, one of the principal activities of every woman just fifty years earlier, had declined to between 10 and 15 percent of all bread consumed; commercial bakeries accounted for the other 85 to 90 percent. Commercial bakeries were also producing more sweet goods than ever, close to 40 percent.[77]

Also in 1950, Royal published a short cooking pamphlet with a long name: *Here Are the Cakes America Loves: Royal Cakes Made with Royal Cream of Tartar Baking Powder*. In spite of the title's emphasis on cakes, the forty pages included recipes for cookies, quick breads, waffles, muffins, biscuits, griddle cakes, frostings, icings, and fillings. For years, Royal had been on the offensive in the baking powder war; now it was all defense. It is clear from this cookbook, with its dreary, asymmetrical cover of plummeting crowns, that Royal was in retreat.

The "double acting" campaign, intensified by the baking powder comparisons in *The Joy of Cooking*, had taken its toll. Royal had to justify itself as single-acting:

it had a "smooth, all-in-one action." It also touted its "steady action." However, Royal faced three insurmountable drawbacks. First, Royal had to be used differently from almost all other baking powders. It had to be added at the end of the mixing, "*with the last addition of flour*," because once Royal came in contact with liquid, the single action was activated. This meant that if the cook added Royal to the batter and it came in contact with liquid, and then the telephone or doorbell rang or a child needed something, the cake would be ruined. Second, the cookbooklet claimed that Royal could be used "in the same amount called for" by other types of baking powder, but its recipes still used larger amounts.[78] Other baking powders used one teaspoon of baking powder per one cup of flour, but this cookbook uses 4 teaspoons of baking power to 1¾ cups of flour.[79] If Royal was in fact now interchangeable with other baking powders, it meant that all previous Royal cookbooks were obsolete. Third, Royal was still vulnerable to overmixing, which is difficult to define. Telling the cook, especially a beginner, that the batter "should not be mixed too long or too vigorously" is a useless instruction. Other instructions were overly detailed yet still incomprehensible: "If you do not beat at a speed of 300 strokes in 2 minutes, add baking powder at last beating period."[80] Previously, confusion had existed between Royal and alum baking powders. By 1950, attempting to use Royal was confusing in and of itself. Clabber Girl and Calumet had the technological and economic high ground. Royal was in an indefensible position.

Royal was contending not only with Clabber Girl, Calumet, KC, and Davis but also with a new type of baking powder created during the war. Byron Smith, of Bangor, Maine, created a baking powder using acid sodium pyrophosphate. He called it Bakewell. It still has a loyal following in parts of New England.

In 1950 Clabber Girl celebrated its centennial with a dinner for 600 employees. A few weeks earlier, Tony Hulman had done something that was in keeping with what Herman Hulman and Anton Sr. had done: he gave Clabber Girl's 280 employees who were members of the Teamsters Union a raise of four cents an hour on top of their new four-month-old contract.[81] At the dinner, employees with more than fifty years of service were honored with "diamond set gold pins and a silver pen."[82]

Clabber Girl also bought two gifts for its one hundredth birthday: Rumford and KC. Rumford had never recovered from the pummeling it had received in the price war and the costs of unionization. In 1940–1941 Rumford had built a modern laboratory in East Providence, but after World War II the company was prevented from further expansion because of a lack of capital. A plan of reorganization was filed in September 1948, and the firm's chemical and baking powder arms were split. Heyden Chemical Corporation in New Jersey purchased Rumford's chemical branch; KC had moved from Chicago to Little Rock, Arkansas, so Clabber Girl acquired its holdings there. At its headquarters in Terre Haute, Clabber Girl was the leading player in the postwar world as baking powder began its second hundred years.

CHAPTER 11

Baking Powder Today

Post–World War II to the Twenty-First Century

Whole Wheat Country Loaf
Active Time: 70 minutes
Total Time: Between 5 and 10 days
—Michael Pollan, *Cooked*, 2013

I kid you not, all you need is some milk, a package o' pudding and a box o' cake mix and you can bake a cake. IT'S SCIENCE.
—Stephanie Wise, *Girl versus Dough*, 2012

Clabber Girl entered the second half of the twentieth century as the uncontested leader in the baking powder industry, with Calumet in second place. Even though they were both alum baking powders and interchangeable in recipes, the nineteenth-century confusion about baking powders persisted in cookbooks until the second half of the twentieth century. Self-rising flour, cake mixes, and commercial bakeries continued to erode the business of baking powder companies in the United States, while American multinational food companies introduced baked goods made with baking powder, especially biscuits, into global markets, increasing American hegemony in worldwide food. Baking powder gained acceptance in countries that had previously been resistant to it and altered traditional foods. In the twenty-first century the moral issues related to bread and the role of women in baking it that had existed in the nineteenth century still lingered, as did the modern versus antimodern debate over food. Just as they had in the seventeenth, eighteenth, and nineteenth centuries, American women in the late twentieth and early twenty-first centuries continued to experiment in their kitchens, looking for ever faster ways to bake and sharing the information in traditional cookbooks and via the new technologies of e-cookbooks and internet blogs.

In prosperous post–World War II America, Royal's market share continued to dwindle. In 1957 Royal published its last cooking pamphlet, *Here Are the Cakes America Loves: Royal Cakes Made with Royal Cream of Tartar Baking Powder*, a reprise of its 1950 cooking pamphlet by the same name.[1] Royal limped along into the 1970s, but then it was gone without fanfare, citing the expense of cream of tartar.[2] In his 1981 cookbook, *The New James Beard*, America's "father of gastronomy" advised home bakers who lamented Royal's passing that they could make their own baking powder in a ratio of two parts cream of tartar to one part baking soda. He also pointed out the two main problems with preindustrial baking powder (and the reasons for Royal's demise): "don't try to store it, though, because it won't keep well"; and cream of tartar baking powder had to be doubled for each recipe.[3]

However, Royal's legacy lingered and there was still chaos in the cookbooks, because there was still no standardization of baking powder. Irma Rombauer's *The Joy of Cooking* was a prime example. It was not until the mid-1960s that sodium aluminum sulfate baking powders like Clabber Girl and Calumet became the standard in *Joy*, but with a qualifying description: "combination, or double-acting, baking powders . . . are the baking powders we specify consistently in this book."[4] Just to make sure there was no misunderstanding, the individual recipes said "double-acting baking powder." Finally, in the 1990s, almost three-quarters of a century after Calumet began to use the phrase "double-acting" in its advertising, *The Joy of Cooking* dropped "double-acting" and simply said, "baking powder."[5] In a break with previous editions, the words "baking powder" did not even appear in the index. The baking powder paradigm shift was complete. Almost 150 years after commercial baking powder production began, sodium aluminum sulfate—"alum"—baking powder, with no further qualifications, was the default.

Ironically, baking powder, which began as an economical way to leaven foods instead of using eggs, butter, or sour milk, became the leavener for foods that were heavy in those ingredients. A comparison of muffins from 1902 to 1997 reveals changes in muffin size, calories, and ingredients, including baking powder. In 1902 a standard muffin, like traditional bread, contained neither sugar nor butter. In the 1920s a weight-loss muffin remained substantially the same as the 1902 muffin. It had no sugar, but it added butter. The muffin for people who wanted to gain weight, called a "Rich Muffin," contained sugar, butter, and eggs. By 1975, what had been a weight-gain muffin in the 1920s was the default muffin. Both the weight-gain muffin in the 1920s and the standard muffin in the 1970s had fewer calories than a low-fat muffin in 1997. The baking powder had to be increased to leaven the additional fat and sugar.[6]

Not only the amounts of the ingredients changed, but so did the method of preparation. In 1975 the ingredients were mixed using the muffin method. The dry ingredients included sugar. The wet ingredients—eggs, butter, and milk—were added to the dry ingredients with a few strokes. By 1997, however, muffins were mixed using the cake technique, where the butter and sugar are "creamed"—whipped to a light, creamy consistency to incorporate air. Then the other ingredients are added. This makes a muffin that is less like bread or a biscuit and more like cake.

What kinds of muffins Americans were eating changed too. In 1904 muffins were made from whole grains: graham flour, oats, rice, rye, and corn. The 1975 edition of The *Joy of Cooking* still had whole-grain muffins, but by 1997 the whole-grain muffins were gone except for bran and corn. New muffins that contained nuts, a doughnut muffin, and a double-chocolate muffin with chocolate chips were added. The 1997 edition contained something else new: toppings that stopped just short of being frosting. One topping was cinnamon-sugar. Another was sugar and almonds. The doughnut muffin sported a sugar, cinnamon, and melted butter topping (see appendix table A-6).

Beginning in the 1950s American women could avoid baking altogether, because new technologies presented easy alternatives. Convenient ready-made desserts like peaches and pound cake, every hostess's friend, tempted homemakers away from baking. The peaches were canned, the pound cake was precooked, frozen, and branded with a memorable ungrammatical slogan: "Nobody doesn't like Sara Lee." Both peaches and pound cake could be topped off with Reddi-Whip, real whipped cream that made use of another new technology, the aerosol can, or Cool Whip, which looked like whipped cream but was made from corn syrup and vegetable oil.

However, the reaction of American housewives to another new product, cake mixes, showed that women were picking and choosing their technologies. For women, the "lovin'" in the Pillsbury slogan "Nothin' says lovin' like somethin' from the oven" meant cake. In the 1930s women had rushed to buy Bisquick to make America's biscuits, quick breads, pancakes, waffles, and muffins. But when Betty Crocker debuted cake mixes in 1947, with Duncan Hines soon following, housewives did not rush to buy them.[7]

This aversion seemed to apply only to layer cakes, which were usually round, frosted, decorated, and served on special occasions, such as birthdays. The cake contests and other competitions that baking powder and flour companies had sponsored to sell their products in the nineteenth century had been effective in teaching women to bake the new kinds of American cakes made with chemical leaveners. Corporate cookbooks with a variety of baking powder cake recipes and multiple styles of frosting, icing, and glazes had made it possible for women to develop personal signature cakes, which then became family traditions. Women took pride in their cakes and earned culinary capital with them. They were not ready to relinquish their cake making.

White Lily and Martha White

> The Southern housewife has always prided herself on the whiteness and flakiness of her biscuit.
>
> —Henrietta Dull, *Southern Cooking*, 1928

Cake mixes were not the only product that cut into the baking powder market. So did self-rising flour with built-in chemical leaveners. Self-rising flours were especially popular in the South. Unlike Minnesota's flour companies, which are named after their founders, like Pillsbury, or to advertise the prizes they have won, like Gold Medal, the South's two major flour companies have names based on race and gender, specifically white womanhood: White Lily, founded in 1883, and Martha White, which started as the Royal Flour Mill in 1899. Both companies developed their own combinations of "soft," low-gluten flours with proprietary baking powder formulas that make Southern biscuits distinctive.[8] Martha White is now owned by the J. M. Smucker Company, which began as a cider mill in Orrville, Ohio, in 1897.[9] Smucker's also owns Crisco shortening, Pillsbury baking mixes and ready-to-spread frostings, and Hungry Jack pancake mixes and syrups.[10] Baking powder is in almost everything but the syrup and the frostings.

Throughout the postwar years, Martha White added convenience and speed to its flour. In the 1950s it created Hot Rize, a self-leavening flour.[11] This baking powder formula is also in Martha White's Cotton Country Cornbread Mix and its other cornbread and muffin mixes: Buttermilk, Yellow, Sweet Yellow, and Sweet Yellow Honey. In the 1960s Martha White introduced Bix Mix, a just-add-water biscuit mix. All three current varieties of the Quick & Easy Biscuit Mix—blueberry, cheese garlic, and cinnamon—contain baking powder. So do the nineteen muffin mixes.[12]

Martha White's advertising capitalized on its location and proximity to country music. In the 1940s the company began to advertise on radio WSM in Nashville—Music City—with "Martha White Biscuit and Cornbread Time." The program aired at 5:45 a.m., when women were preparing breakfast.[13] In 1948, with a marketing budget of twenty-five dollars per week, Martha White began to sponsor the Grand Ole Opry radio show on Saturday nights.[14] The Martha White name was featured on the stage backdrop, above the title of the show. The bluegrass duo of guitarist/mandolin player Lester Flatt and banjo picker Earl Scruggs, known as the Foggy Mountain Boys, recorded the company's jingle, "You Bake Right with Martha White":

> Now you bake right (uh-huh) with Martha White (yes, ma'am)
> Goodness gracious, good and light, Martha White
> For the finest biscuits, cakes and pies,
> Get Martha White self-rising flour
> The one all purpose flour,

Martha White self-rising flour's got Hot Rize
For the finest cornbread you can bake,
Get Martha White self-rising meal
The one all purpose meal
Martha White self-rising meal
For goodness' sake.[15]

The Grand Ole Opry, Martha White's biscuits, and their slogan since 1944, "Goodness Gracious, It's Good!" were parodied on National Public Radio's *Prairie Home Companion*, set in Minnesota.[16] Powdermilk Biscuits was one of the fictitious sponsors of the musical portion of the show, along with Mournful Oatmeal ("Calvinism in a box"), the American Duct Tape Council, and Earl's Academy of Accents.[17]

Host Garrison Keillor began with the biscuit advertising blurb:

> Powdermilk Biscuits, made from whole wheat, that gives shy persons the will to do what needs to be done. And you know there are times when one must do it. And you never know until you come right to the brink and then you see what you've got to do, and—Powdermilk Biscuits. Whole wheat.
>
> Heavens, they're tasty. And expeditious.[18]

Keillor often added that Powdermilk Biscuits were "made from whole wheat raised by Norwegian bachelor farmers, so you know they're not only good for you, but also pure, mostly."[19]

The International Biscuit Festival

America's—and especially the South's—love affair with baking powder biscuits has grown stronger in the twenty-first century. In 2009, Knoxville, Tennessee, hosted the first International Biscuit Festival. By 2014, attendance had grown to more than twenty thousand.[20] Bakers competed in four categories: Sweet Biscuits, Savory Biscuits, Special Biscuits, and Student Biscuits. The winners received cash prizes of up to three hundred dollars.[21] The entries, which reflect industrial advances and ethnic diversity, illustrate how far baking powder biscuits have come since they were created in the early nineteenth century and how versatile America's iconic bread is.

The biscuits were made with many types and brands of flour, leavening, and shortening: White Lily flour, Southern Biscuit self-rising flour, all-purpose flour, cornmeal, and biscuit mixes. Shortenings included cream, sour cream, cream cheese, butter, buttermilk, powdered milk, skim milk, evaporated milk, lard, mayonnaise, yogurt, vegetable oil, margarine, Crisco, and Cool Whip.

Additions to the biscuits included proteins, fruits, vegetables, herbs, spices, nuts, cheeses, olives, soda pop, chocolate chips, and cayenne pepper. Recognizably

Southern biscuits incorporated sweet potatoes, fried green tomatoes, bacon, sausage, ham, pecans, or blackberry moonshine in the biscuit itself or as accompaniments. Baking powder biscuits also easily adapted to ingredients from multiple ethnic groups, like the prizewinning "Korean Kimchi Bulgolgi Biscuits" and biscuits that used chipotle mayonnaise, adobo sauce, or jalapeño cilantro *verde* sauce.

Bakers used multiple culinary techniques to transform biscuits after baking. Biscuits were battered and turned into French toast; they were also used for sandwiches filled with catfish, beef, pork, brandied onion rings, or beer-battered onions. They were topped with caramel sauce, maple syrup, confectioners' sugar, and glazes made with coffee, oranges, apple cider, peanut butter, or rum or served with a side of honey-mustard dipping sauce.[22]

In spite of all the new notes that participants in the International Biscuit Festival rang on baking powder biscuits, one of the most iconic desserts across the South is still peach cobbler, with its baking powder biscuit topping. Peach cobbler is another "traditional" dish for which the tradition stretches back only as far as the invention of baking powder.

ADRIAN LIPSCOMBE'S PEACH COBBLER

Uptowne Café, LaCrosse, Wisconsin, hello@uptownecafe.com

Filling

1½ tablespoons cornstarch
¼–⅓ cup brown sugar, depending on how sweet the peaches are
½ cup water
4 cups fresh peaches, peeled and sliced
1 tablespoon butter
1 tablespoon lemon juice
1 tablespoon cinnamon

Topping

3 cups flour, sifted
1 tablespoon baking powder
1 teaspoon salt
⅓ cup vegetable shortening or butter
1 cup milk

Heat the oven to 425 degrees F. (220 C).

Filling

1. In a large saucepan, mix cornstarch and brown sugar with water. Add peaches and cook until mixture is thickened, about 15 minutes, stirring occasionally.
2. Stir in butter, lemon juice, and cinnamon.
3. Pour into an 8-inch round baking dish.

Topping

1. Mix sifted flour, baking powder and salt in a bowl.
2. Cut in shortening until mixture looks like coarse meal.
3. Add milk and stir just enough to hold dough together.
4. Drop the batter topping by spoonfuls onto the peach mixture, and spread evenly.
5. Bake until topping is golden, about 30 minutes. Serve warm.

In 2015 the baking powder biscuit finally rose above its humble beginnings and achieved superstardom. It was ready for its close-up in a food porn commercial. For years, fast food company Carl's Jr. had been sexing up its burger ads with bikini babes biting into burgers that dripped juice all over them. But in 2015 Carl's Jr. cut out the women and personified the biscuit as the star. Tendrils of steam sway seductively as the camera pans down to the hot body of the biscuit, which fills the screen. Manicured male fingers, thumbs splayed at the bottom, gently pry the sides apart at an angle, revealing the soft, tender inside. A male voice announces, "This is made from scratch"—a virgin biscuit. Cut to the biscuit, now stuffed with pink meat, as it bounces down onto the counter.

This commercial is the opposite of the text-dense product information that characterized nineteenth-century advertising. No wall-to-wall, top-to-bottom scientific information about why the viewer should eat this biscuit. No testimonials from scientists, physicians, or eminent personages. This baking powder biscuit commercial carries the pleasure principle to an extreme. It is all hedonism and orgasm.

Baking Powder Goes Global

By the twenty-first century, some of the most iconic foods sold by the world's largest quick service restaurants (QSRs) were leavened with baking powder. A new type of baking powder, SAP—sodium acid pyrophosphate—made volume production of baked goods possible. This slow-acting chemical is used primarily in "slack-dough products"—that is, doughnuts and cakes—where there is a delay from the time the machine mixes the batter until it goes into the oven.[23]

In 2010, QSRs received $174 billion of the $361 billion that Americans spent in restaurants.[24] Many of the foods sold in QSRs contain baking powder. McDonald's, the world's largest global restaurant chain, introduced the Egg McMuffin in 1972. One of the leavening agents in the muffin is monocalcium phosphate, now mined, but which Eben Horsford made so painstakingly from burned bones in 1856. It is on the breakfast menu, along with thirteen other biscuit, McMuffin, or griddle sandwiches, most of which contain baking powder.[25] The second largest global restaurant chain, Kentucky Fried Chicken, with more than twenty thousand outlets in 125 countries and territories, is known for its buttermilk baking powder biscuits.[26] Dunkin' Donuts also uses baking powder to leaven its cake-style donuts.[27]

Breakfast has become so popular that some restaurants serve it all day, and they rely heavily on baking powder. In 1958 the International House of Pancakes (IHOP) opened in Toluca Lake, California.[28] In 2014 the menu had sixteen different pancake types and combinations, along with biscuits and waffles that used baking powder. Denny's, too, promotes its all-day breakfast menu of baking powder–leavened breadstuffs. Starbucks and other coffee vendors sell baking powder–leavened cakes, muffins, and cookies. In 1977, against the advice of friends and business advisors, a Northern California woman named Debbie Fields opened a store that sold only cookies. Cookies are another American original, first mentioned in 1796 by American orphan Amelia Simmons, who leavened them with the original baking powder, pearlash.

At the end of the twentieth century, aging baby boomers revisited their childhoods through cupcakes. In the 1950s, cupcakes—often made from Jiffy cupcake mix, which contained chemical leaveners, and topped with buttercream frosting and sprinkles—became standards for children's birthday parties and school celebrations. They were less messy than cake, and no knife was necessary to cut them, because they were already in individual portions. They were sanitary because the bottom was in a paper cupcake liner, so the cakes were not touched, only the paper. They were festive because the cupcake papers were fluted and came in many pastel colors. On July 9, 2000, the cupcake craze was officially reignited, but as sexy adult food, when the HBO hit television series *Sex and the City* featured its stars eating cupcakes on a bench outside Magnolia Bakery in New York City.[29]

In 2005, Sprinkles, the world's first cupcakes-only store, opened in Beverly Hills, California. Sprinkles' founder, entrepreneur and pastry chef Candace Nelson, pioneered three new ways to sell cupcakes. On June 13, 2010, the reality television competition show *Cupcake Wars* debuted, with Nelson as a judge.[30] In 2012 Sprinkles installed the world's first cupcake vending machine on the street outside its Beverly Hills store. The frosting-pink cupcake ATM satisfies cupcake cravings 24/7. And patrons can track the chocolate-brown Sprinklesmobile, the first cupcake truck in the world, on Twitter. By 2017 Sprinkles had twenty-four stores in nine states and Washington, D.C., and was growing. It had also expanded into ice cream, with a red velvet waffle cone.[31]

Baking powder also spread to France and Italy, two countries that had adamantly resisted it earlier. In 1927 the cookbook that influenced Julia Child, *Le Livre de Cuisine de Mme. E. Saint-Ange*, had two recipes for plum cake that used "English powder"—"*levure en poudre anglaise* (baking powder)." Cookbooks that adapted French recipes for American kitchens also added baking powder. In 1961 Julia Child included baking powder—she made sure to specify double acting—in her recipe for the French standard tart crust, *pâte sablée*.[32] In 1975, when Philip and

The world's first cupcake ATM vending machine, outside of Sprinkles Cupcakes in Beverly Hills, California.

Mary Hyman "revised and adapted" the great French pastry chef Gaston Lenôtre's recipes for American cooks—Gâteau Basque and *génoise*, the basic French cake—they were leavened with baking powder.[33] By the end of the twentieth century, the *Larousse Gastronomique* recipe for madeleines, the small, shell-shaped tea cakes that originated in the eighteenth century and that Marcel Proust made famous in the nineteenth, contained baking powder.

Italian desserts, too, now routinely use baking powder. In 2009 the Accademia Italiana della Cucina (the Italian Academy of Cuisine) published *La Cucina: The Regional Cooking of Italy*. The purpose of the book, which contains "2,000 authentic recipes," was to memorialize traditional regional Italian cooking before it disappeared. Too late. Some desserts are still leavened in traditional ways, with yeast, wine, or egg whites. However, baking powder is now in many of Italy's iconic desserts: biscotti, the crust for ricotta cheesecake, and the St. Joseph's Day fritters known as *zeppole*. In spite of the Accademia's best efforts, Italian millennial women are on internet sites like Tribù Golosa (Tasty Tribe), where they can read the "Top Blog"; enter a "contest"; and post their versions of American recipes new to their culture for pancakes, muffins, cookies, and waffles, and other baking powder–leavened foods.[34] What appeals to them about these recipes is that they are "*veloce*" and "*facile*," quick and easy—the same qualities that drew American women to baking powder 150 years ago. An Italian brand of phosphate baking powder is Paneangeli—angel bread. It is sold not in cans, as it is in America, but in individual premeasured packets.

Baking powder has permeated the cooking of other cuisines. The extruded Mexican cruller-like dessert, churros, are now leavened with baking powder. Classic Russian foods like blini, the small crêpes eaten with caviar, are leavened with baking powder. So is honey cake, "among the oldest known Russian cakes."[35] Even Belgium's national spice cookie, Speculoos, contains baking powder.[36]

Although it had faded away from American shelves in the postwar years, Royal gained popularity in foreign countries. In Spain, baking powder is known as *polvo Royal*(Royal Powder) or *levadura Royal* (Royal leavening)—that is, not yeast. It is sold in individual packets with recipes for Spanish foods that have traditionally been leavened with yeast, such as empanadas. The allure, as always with baking powder, is speed.[37]

In South Africa, Kraft continues to distribute Royal Baking Powder through its subsidiary Mondelez South Africa.[38] Headquartered in Johannesburg, Royal runs contests that appeal to the black community, still dependent in many areas on *spazas*, small local convenience stores sometimes operated in homes. Spazas date from pre-apartheid times when there were few services in black communities; the word is slang for "just getting by."[39] In one contest, the prize was a *spazatainer*, similar to a small prefabricated shed, in bright red with the Royal logo on it. In another contest the prize was a bright red mobile kitchen with the Royal logo.[40]

The Burden of Bread, Redux

While other countries embraced baking powder and modernity, the antimodernist movement revived in the United States. Modern muckraking about food began in earnest 2001 with the publication of Eric Schlosser's book *Fast Food Nation: The Dark Side of the All-American Meal*, followed by Marion Nestle's *Food Politics: How the Food Industry Influences Nutrition and Health* in 2002, and Michael Pollan's *The Omnivore's Dilemma: A Natural History of Four Meals* in 2006. All were best sellers and award winners, and all dealt with many of the same issues that concerned Americans during the baking powder wars in the nineteenth century: the addition of chemicals to food, the connections between business and the food supply, how businesses use the media to advertise, and the role of government in regulating business and food. All were also devoutly anti–big food business.

Michael Moss's 2013 Pulitzer Prize–winning muckraking book, *Salt Sugar Fat: How the Food Giants Hooked Us*, faulted food corporations for "capitalizing on society's hunger for faster, more convenient food."[41] Moss states that Americans are unable to resist these foods because food scientists in the employ of multinational corporations have spent millions, perhaps billions, of dollars to make sure that each food hits its "bliss point"—the combination of flavors that makes it irresistible. Moss views consumers as passive dupes at the mercy of corporate America.[42]

However, as the history of baking powder shows, consumer demand has been driving the corporate train. Baking powder began in the eighteenth century with women experimenting in their own kitchens to find shortcuts. Corporate America did not force housewives to stampede to the supermarket in 1931 to buy Bisquick.

Women bought Bisquick because they liked what they read on the package—"90 SECONDS from package to oven"—and because the product delivered what it promised.[43]

Consumer demand caused Bisquick and its imitators to proliferate and continue to do so. By 2015 there were multiple Bisquick mixes: Bisquick Original; Bisquick Complete Mix: Buttermilk Biscuits, Cheese Garlic Biscuits, Honey Butter Biscuits, and Three Cheese Biscuits; Bisquick Pancakes Complete; and Bisquick Heart Smart Pancake and Baking Mix. That was still too slow for Americans, so Bisquick came out with Bisquick Shake 'N Pour.[44]

Consumer demand also kept Jiffy baking mix selling strong. In 2015 Jiffy celebrated its eighty-fifth anniversary. It had a lot to celebrate: it controlled approximately 65 percent of the prepared muffin-mix market and 90 percent of the corn muffin-mix market.[45] Like Clabber Girl, Jiffy is still family owned and run by the fourth generation. Under president and CEO Howard Samuel "Howdy" Holmes, the company had expanded from selling to supermarkets to institutional markets. On October 29, 2015, Howdy's grandmother and founder of Jiffy, Mabel White Holmes, was inducted into the Michigan Women's Hall of Fame.[46]

Much of the contemporary antimodernist movement still centers on the mythology of bread baking. In the 1960s women who tried to live up to an impossible image of the ideal housewife, what Betty Friedan termed "the feminine mystique," baked their own bread.[47] In 2013 Michael Pollan, professor of journalism at the University of California, Berkeley, wrote that cooking "is the most important thing an ordinary person can do to help reform the American food system, to make it healthier and more sustainable."[48] In one eighty-six-page chapter, Pollan explains how he learned the joys of bread baking under the tutelage of a male professional baker. He then provides a recipe that is six pages long, takes five to ten days to prepare, and results in two loaves of bread. Pollan uses a starter that captures natural yeasts from the air and has to be fed every day, but he still finds it necessary to add commercial yeast, rapid-rise or instant, "as an insurance policy." Both rapid-rise and instant yeast are laboratory products, creations of recombinant DNA technology. Pollan's ingredients, precise and scientific, are in grams, for which a modern device, a digital scale, is necessary. However, the variables for Pollan's bread are the same as they were for women who used homemade yeast thousands of years ago: "the ambient temperature and the vigor of your starter."[49] These were the two most important variables that commercial baking powder replaced with certainties.

Pollan's message about the virtues of bread baking could have come from Sylvester Graham's 1837 *Treatise on Bread and Bread Making* and is presented with the same missionary zeal. What Pollan celebrates about bread baking is what Eben Horsford, in his 1861 *The Theory and Art of Bread-Making: A New Process without the Use of Ferment*, saw as a problem and for which he presented baking powder as a solution. What Pollan sees as a positive reversal of industrialism is a burden that nineteenth-century women were elated that baking powder, industrialization, and corporate America had lifted from them.

On the other side of this argument is food historian Rachel Laudan, who calls the "moral and political crusade" against industrialized food "Culinary Luddism." Her 2013 book, *Cuisine and Empire*, expanded on her 2001 seminal essay, "A Plea for Culinary Modernism: Why We Should Love New, Fast, Processed Food."[50] In both article and book, Laudan defends industrialized food as being of better quality, and available to more people, than preindustrial food. She echoes Andrew Carnegie's 1889 statement about the beneficial radical changes that industrialization made in the standard of living: "The poor enjoy what the rich could not before afford. What were the luxuries have become the necessaries of life."[51] Many of those luxurious necessities, like cake and biscuits, are products of the baking powder revolution.

Laudan also points out that most foods that we consider traditional are relatively new and are the products of industrialization. For example, many of the recipes that we now consider traditional, typically American, and comfort food were hailed as new in Royal Baking Powder's cook booklet *The Royal Baker and Pastry Cook* in 1878—not so long ago.[52]

Eating Like the Mythical Grandmother

Hand-in-hand with the mythology of bread and tradition is the culinary nostalgia for a mythical grandmother. The words "country," "grandmother," and "old-fashioned" in any combination have always had tremendous marketing value for Americans. Americans' obsession with Grandma's cooking—usually baking, and almost always something made with baking powder—stretches back centuries. "Grandma's Old Fashioned Country Gravy" is White Lily flour's recipe for biscuits and gravy. In 1942, in *Cross Creek Cookery*, Marjorie Kinnan Rawlings wrote, "Any child who does not have a country grandmother who keeps a cooky jar is . . . to be pitied."[53] A 1939 Southern cookbook has a recipe for "Grandmother's Caramels."[54] The 1935 General Mills cookbook, *How to Take a Trick a Day with Bisquick*, has a recipe and a picture of "the old-fashioned chicken pie that made grandmother's cooking famous," updated and made easily with Bisquick. The 1876 *Centennial Buckeye Cook Book* recipe for "South Carolina Biscuits" calls for shaping them by hand "as our grandmothers used to do."[55] The popular Christmas song, "Over the river and through the woods / To Grandmother's house we go," began as a poem written by Lydia Maria Child in 1844.[56]

However, a study of American consumption habits in 1950 noted that "despite the frequently heard preference for 'grandma's old-fashioned baking,' consumers apparently showed no indication of reverting back to home-made baked foods."[57] Americans were—and are—increasingly content to let business do the baking.

Nevertheless, in 2008 Michael Pollan created his "Great Grandmother Rule": "Don't eat anything your great grandmother wouldn't recognize as food."[58] Great Grandma was eating refined white flour, sugar, and cake made with baking powder when she wasn't eating commercial products like Oreos, Aunt Jemima pancakes, and graham crackers.

Along with nostalgia for the mythical grandmother is the male lament for lost female cooking skills. In a September 2013 article, "The Joy of Cooking," food historian Joseph Campesi listed what he considers the detrimental aspects of fast food: "Fast Food does not invite cooperation in getting a meal ready, allowing us to make do without the once-shared tasks of cutting and chopping, stirring and mixing, heating and baking. . . . Individuals are not called upon to possess or develop any complex knowledge or skills, to be productive or active in any way. One need not know how to prepare food, how to cut and chop, how to knead and fold, how to roast and bake."[59] These are skills that men previously neither had nor considered valuable in the home, but now that women are breaking free from the constraints of the kitchen, men are objecting. Meal preparation in the home has traditionally been performed by women, often with little or no help, in spite of Campesi's "once-shared tasks." Often, men's knowledge of these female skills was only as voyeurs and beneficiaries. Meanwhile, the women who have had to do the work welcome shortcuts like baking powder. Campesi also does not consider what new skills, more valuable and marketable, might have taken the place of cooking skills, or what activities with family might replace the time taken by laborious baking. Women's rebuttal to this male nostalgia is on the internet.

> "Yes people rant about stop being lazy and make something from scratch. Sorry, I'm busy."—Edna, in an Amazon review for Cathy Mitchell's *Quick & Easy Dump Cakes and More*

Michael Pollan, Joseph Campesi, and other antimodernists are concerned with process, which can be lengthy. Modernists and the women who do the cooking are concerned with results. American women continue to experiment in their kitchens as they did more than two hundred years ago, still searching for shortcuts, especially for breadstuffs and cake. Invariably, they find them. Minnesota blogger Stephanie Wise, aka Girl Versus Dough, explains the dump cake: "It's not called a dump cake for any reasons other than good ones—like dumping a bunch of yummy ingredients together into a pan and letting them all bake together and do their thang [*sic*] and before you know it, finish off as a legitimate cake."[60] The crucial ingredient that makes dump cakes possible is baking powder.

Dump cakes are full-size cakes, suitable for families but too large for the increasing number of smaller and single-person households. Enter mug cakes, which are one- or two-portion cakes mixed in a mug and cooked in the microwave for approximately one minute. Baking powder is usually the leavening ingredient.

More than two hundred years after Amelia Simmons included pearlash in three recipes for gingerbread and introduced "Soft Gingerbread Baked in Pans," there is "Gingerbread Mug Cake," leavened with pearlash's chemical descendant, baking powder.

GINGERBREAD MUG CAKE

Makes 1 serving.

1 tablespoon butter
1 tablespoon molasses
3 tablespoons water
¼ cup all purpose flour
¼ teaspoon baking powder
¼ teaspoon powdered ginger
¼ teaspoon ground cinnamon
1 teaspoon white sugar
1 teaspoon brown sugar (or 1 more teaspoon white sugar)

1. Melt the butter in a mug in the microwave.
2. Stir the molasses and water into the butter.
3. Add the remaining ingredients to the mug. Stir well to combine.
4. Microwave on High for 1 minute.

On the internet, American women write about the new shortcuts they create for baked goods, which almost always contain baking powder, in another new kind of cookbook they have pioneered: the e-cookbook. The need for women to connect and share kitchen shortcuts has not changed; only the technology of connection has. Just as cookbook writers from the eighteenth and nineteenth centuries—Amelia Simmons, Sara Josepha Hale, Lydia Maria Child, Mary Randolph, Eliza Leslie, Catharine Beecher—gave life advice and opinions along with their recipes, bloggers do this now on the internet. The sweet and savory recipes in Stephanie Wise's e-cookbook *Quick Bread Love*—"No-Yeast Cinnamon Rolls," "Chocolate Hazelnut Swirl Coffee Cake," "Sun-Dried Tomato Spinach Bread"—are innovations made possible by the baking powder quick breads that American women created through kitchen experiments and first wrote about two hundred years ago.[61]

E-cookbooks differ from print cookbooks in several ways. Like cookbooks from earlier centuries, e-cookbooks use testimonials for promotion. But these testimonials are different. Traditional publishers of hardcover cookbooks quote literary authorities such as *Publisher's Weekly*, *Library Journal*, and culinary authorities. The e-cookbook, however, has testimonials from smiling young women, strangers with no particular credentials, which make them just like the e-book's target market. They praise e-books for their easy-to-use recipes and beautiful pictures. The medium makes these testimonials as valid as, if not more valid than, anonymous traditional journals. Also, e-cookbooks can be delivered to a computer or mobile device and come with ancillaries that traditional cookbooks do not, such as video tutorials and printable recipe cards.[62]

In the twenty-first century, technology reduced the time for bread and cake baking even more, from hours or minutes to seconds. It did this by cutting out yeast and baking powder completely. Placing the batter in an aerosol can forces air bubbles into the batter and aerates it instantly. No rising time is necessary, because there is no chemical reaction with liquid or heat. On January 18, 2010, celebrity chef Michael Voltaggio uploaded a video to YouTube and demonstrated how to make the French sweet bread brioche using aerosol technology. He combined eggs, butter, milk, flour, sugar, and salt in a blender, put the batter into an aerosol container, and sprayed it into a paper cup. After forty-five seconds in the microwave and another thirty to sixty seconds to finish cooking outside of the microwave, fresh bread was ready to eat.[63] In 2014 two Harvard students patented Spray Cake, the aerosol process applied to cake making. It produces a light, fluffy cake in sixty seconds in the microwave.[64]

Conclusion

Baking powder created a true scientific revolution. It caused a paradigm shift and then vanished from consciousness. Americans today are used to eating foods leavened with baking powder, although they are unaware of it, as they are unaware that there was ever any question about it. Once the source of so much controversy, lawsuits, and bitter contention that ranged from every American kitchen to the White House and the United States Supreme Court, baking powder is now an invisible ingredient—an *indispensable* invisible ingredient. It is an American invention, and it was crucial in creating a uniquely American cuisine that has spread throughout the world.

Today there are two major baking powder companies in the United States: Clabber Girl and Calumet. Both contain sodium aluminum sulfate, as they did when they were created in the nineteenth century. Calumet's market share has stagnated at about one-third, the same as eighty years ago before the price wars of the Depression. Then it was the market leader; now it is the minority, dwarfed by Clabber Girl's nearly two-thirds' share. Printed on the label of every can of Clabber Girl baking powder is "America's Leading Brand."

In 2008, Kraft, which owns Calumet, introduced a nonaluminum baking powder through another of its products, Argo cornstarch. The acid ingredients in Argo Double Acting Baking Powder are monocalcium phosphate, Horsford's invention from approximately 150 years ago, and sodium acid pyrophosphate. It is packaged in a "stay-fresh container," a square plastic tub with a screw-top lid, with a "leveling edge" and is sold at Walmart and Sam's Club.[65]

In 2002 Clabber Girl acquired Hoboken, New Jersey–based Davis Baking Powder. In 2010 Clabber Girl owned Royal for domestic U.S. distribution in the

Hispanic market. Royal's ingredients illustrate the complete collapse of the baking powder wars and reflect market standardization. Cream of tartar, the focus of nearly a century of advertising and the basis of lawsuits, is completely missing in the new Royal. The formula contains the same sodium aluminum sulfate found in Calumet and Clabber Girl and the same monocalcium phosphate invented by Eben Horsford. The Royal label reads "Double Acting."

Epilogue: Baking Powder Legacies

All of the families connected with major baking powder businesses left cultural and financial legacies. Like baking powder itself, many of the legacies are connected to speed. The men responsible for the success of baking powder brought their vision, imagination, and business expertise to their hobbies and avocations.

The Ziegler family is well known in yacht racing. They have also donated heavily to charities, especially those for the blind. This continues a tradition begun by William Ziegler Sr.'s wife. Electa Matilda Ziegler (1841–1932), whose son from a previous marriage was blind, began the E. Matilda Ziegler Foundation for the Blind, Inc., in New York City in 1907.[66] It is now based in Darien, Connecticut. The family has also held national and international positions in charities for the blind. In 2011 the fourth generation of the Ziegler family established the William Ziegler III Professorship in Vision Research at the Yale School of Medicine.[67]

The Wright family, too, was fascinated by speed. In 1924 William Monroe Wright bought a Kentucky thoroughbred farm and named it after his baking powder, Calumet Farm. Warren Wright railed against his father's extravagance. However, the purchase did not damage the family finances. When William Wright died in 1931, he left fifty-five million dollars of his sixty-million-dollar estate to his son.[68] Warren decided that he would run the farm as a hands-on owner and bring modern business practices to horse breeding and racing. He was determined to make Calumet Farm at least break even, and he succeeded phenomenally. In 1941 Calumet Farm led United States stables in earnings, with almost one million dollars. The farm continued to be the leader in thoroughbred racing even after Warren Wright's death in 1950. In the half century between 1941 and 1991, Calumet Farm produced nine Kentucky Derby winners. Two of them, Whirlaway and Citation, went on to win the Triple Crown. In 1990 Calumet Farm owned Alydar, who sired more winning offspring than any other horse. One of Alydar's progeny that earned more than a million dollars was named Clabber Girl.[69] The farm's decline into mismanagement, bankruptcy, and criminality in the 1990s is documented in *Wild Ride: The Rise and Tragic Fall of Calumet Farm, Inc., America's Premier Racing Dynasty*, by Ann Hagedorn Auerbach.[70]

Clabber Girl baking powder and the Hulman family have journeyed far. At the end of the nineteenth century, Clabber baking powder was just a sideline, a small part of the thriving grocery business, which was itself part of the larger thriving urban economy of Terre Haute. In the twentieth century, Clabber Girl continued to grow as the Hulman & Co. general store and Terre Haute waned. On March 20, 1963, a major fire burned down much of the business district on Wabash Avenue, close to the Hulman store.[71]

In the twenty-first century, Terre Haute, like many other towns in America, is reinventing itself. The Clabber Girl Corporation is a major industry in Terre Haute. In October 2006 Clabber Girl did something they had not done since the 1920s: broke ground on a new manufacturing facility, seventy thousand square feet, which created forty-two new jobs.[72] The Hulman family still donates to many local charities. They gave such a substantial endowment to the Rose Polytechnic Institute in Terre Haute that the school changed its name to the Rose-Hulman Polytechnic Institute. In December 1973 the ten thousand–seat Hulman Center auditorium at Indiana State University in Terre Haute opened. A one-hundred-thousand-dollar gift to the city from Tony Hulman purchased the 640 acres that became Hulman Field Airport on October 3, 1944, now Terre Haute International Airport–Hulman Field.[73]

The five-story building that opened with such fanfare in 1893 still stands at the corner of Ninth Street and Wabash Avenue, two blocks from the Crossroads of America. The trains are still nearby, but now so are U.S. Highway 40 and Interstate 70. The corporate offices are still upstairs. On the ground floor, where customers once stood on wide oak planks and lined up at carved counters, there are now a museum and a café that serves freshly made baking powder biscuits.

Tony Hulman died in 1977, but all his life he continued his family's tradition of speed-seeking pastimes. His father and uncle had both been state bicycle-racing champions when bicycles were new. His grandfather Herman had treated five thousand guests to harness racing at the opening of the new store in 1893. Tony had been a national track and football star.

In 1945 Tony paid the enormous sum of $750,000 for a run-down car race track that he bought from World War I flying ace Eddie Rickenbacker. People thought Tony was insane. A real estate developer told Tony the track was a tear-down. His mother agreed. His friends were sure he was going to lose the family fortune on this money pit. That was because they couldn't see what Tony saw. He looked past the track that was gapped and cracked, with weeds sprouting out of it, and saw solid new grandstands where ramshackle wooden ones stood.

Located seventy-five miles from Terre Haute, the Indianapolis Motor Speedway had begun in 1909 with a disastrous tire-shredding crushed stone and tar track, but when it was resurfaced with ten-pound bricks set in sand and mortar later that year, it earned the nickname it still has: "The Brickyard." In 1911 the first Indianapolis 500 race was held. But after initial excitement, car racing declined for economic reasons, and nobody missed the track during World War II when

Tony Hulman in an Indy race car. Courtesy of the Vigo County Public Library Community Archives.

the federal government issued a ban on auto racing in 1942 and rationed gas and tires.

Tony was determined to make the speedway self-sufficient. To generate cash flow, he sold the track's concessions for $1 million for ten years and used $250,000 to finance repairs to the track. Tony was an advertising genius and made sure that on opening day in May 1946, celebrities were in attendance: members of the Firestone Tire family, Henry Ford II, boxing champion Jack Dempsey, and Lt. Gen. Jimmy Doolittle, who had led the air raid bombing of Tokyo in 1942.

Still, Tony worried that nobody was going to pay ten dollars for a ticket, even if qualifying speed did start at 115 mph. On race day, streets in Indianapolis were clogged with two hundred thousand people waiting to get into the speedway. After that, Tony ignored Rickenbacker's advice to keep the prize money modest, capped at seventy-five thousand dollars. Instead, Tony Hulman turned the Indy 500 into the highest-paying car race in the world.[74] One racer, who won Rookie of the Year in 1979 and drove in five more Indy 500s, was "Howdy" Holmes, of Jiffy baking mix, which contained baking powder.[75] Like baking powder, the Indy 500 is uniquely American. European car racing is street racing; American racing is on a track.

Tony Hulman resurrected the Indianapolis Motor Speedway. He passed that legacy on to his family and to all Americans. On Memorial Day weekend, the woman who says those iconic American words, "Gentlemen, start your engines!" is Tony Hulman's daughter, Mari Hulman George. Just as Yankee Stadium is the house that Babe Ruth built, the Indianapolis Motor Speedway is the house that Clabber Girl built.

Think about that the next time you bite into a biscuit.

Glossary

albumen: the protein part of the egg, the white; added to baking powders, causes a foaming action; cause of controversy especially for Calumet Baking Powder in the early twentieth century.

alkali: a salt; in chemistry, a base, as opposed to an acid. Baking soda is an alkali.

alum: an astringent acid; incorrectly used by cream of tartar baking powders to refer to sodium aluminum sulfate.

ammonia, aka *ammonium carbonate*, *baker's ammonia*, and *hartshorn*: an acrid substance that smells like urine but has leavening properties.

baker's ammonia: see *ammonia*.

baking soda, aka *sodium bicarbonate*: white alkaline mineral with leavening properties.

cornstarch: the buffer ingredient added to prevent baking soda and other ingredients from combining too soon.

cream of tartar: an acid by-product of wine making; used in Royal Baking Powder.

"*double acting*": refers to the rising action of baking powder first with liquid, then with heat.

emptins: the sludgy yeast left over from home beer brewing; used as a leavener for breadstuffs until the mid-nineteenth century.

hartshorn, aka *ammonium carbonate*, and *baker's ammonia*: pre–baking powder European chemical leavener, especially in Germany.

monocalcium phosphate: the basis for Rumford Baking Powder; originally made from burned bones; later mined.

muriatic acid: hydrochloric acid; used as an early attempt at leavening.

nixtamalization: the Aztec process of adding vegetable ash, which contains B vitamins, to corn.

pearlash, pearl ash, pearl-ash: refined vegetable ash (potash) used as an early leavener.

potassium carbonate: *not* the same as sodium bicarbonate, but also a mineral leavener, and like sodium bicarbonate, also called saleratus, which created confusion.

sal volatile: smelling salts; used as an early leavener.

saleratus, sometimes *sal aeratus*: literally, aerated salt; used to refer to potassium carbonate and baking soda, which created confusion.

sodium aluminum sulfate: the acid leavening agent in most modern commercial baking powders such as Clabber Girl and Calumet.

sodium bicarbonate, aka *baking soda*: white alkaline mineral with leavening properties.

yeast: fungi used beginning thousands of years ago to leaven beer and bread.

Appendix: Baking Powder Use in Cookbooks

Table A-1. Types of Recipes, 1796—Amelia Simmons, *American Cookery*

	1st ed., Hartford	2nd ed., Albany
Meat	18	30
Fish	no separate section	6
Pies	6 savory, 3 sweet	8 savory, 5 sweet
Custards	5	5
Tarts	7	6
Puddings	25	31
Pastes	9	9
Syllabubs (and Creams)	7	7
Cakes	36	46
Pickles and Preserves	30	28
Emptins	1	—
Total Desserts/Sweets	92	103
Total Protein/Savory	24	44

Table A-2. Types of Leavening, 1796–1858

	1796 Amelia Simmons, *American Cookery*	1858 Catharine Beecher, *Miss Beecher's Domestic Receipt-Book*
Yeast	8	34
Emptins	15	—
Pearlash	7	7
Saleratus	—	37
Sal Volatile	—	8
Super Carbonate of Soda	—	5
Cream of Tartar + Alum	—	2
Tartaric Acid	—	1
Total Yeast / Emptins	23	34
Total Chemical	7	62

Table A-3. Types of Leavening, 1876, *National Cookery Book*

Leavener	Number of Recipes	Percentage
Yeast	39	37
Soda	28	27
Cream of Tartar + Soda	14	13
Saleratus	10	9.5
Baking Powder	9	8.6
Yeast Powder	4	3.8
Cream of Tartar + Saleratus	1	.95
Total Leaveners	105	
Total Non-Yeast Leaveners	66	63

Table A-4. Types of Leavening, 1876, *Centennial Buckeye Cook Book*

(Includes 1 recipe for cream of tartar + saleratus, and 2 recipes for cream of tartar or baking powder + soda)

Leavener	Bread	Percent	Cake, etc.	Percent
Yeast	30	.4	5	.03
Yeast + Soda	—		4	.02
Soda	27	.36	42	30
Cream of Tartar + Soda	12	.16	32	23
Cream of Tartar	2	.02	3	.02
Baking Powder	2	.02	46	33
Saleratus	1	.01	2	.01
Yeast Powder	—		2	.01
Clabbered Milk	1	.01	—	
Alum	—		3	.02
Total Leaveners	75		139	
Total Non-Yeast Leaveners	45	60	134	96

Source: *Centennial Buckeye Cook Book*, ix.

Table A-5. Types of Leavening, 1887, *White House Cook Book*

Ingredient	Bread	Percent	Biscuits, Rolls, Muffins	Percent	Cake	Percent	Total	Percent
Baking Powder	3	14	37	45	30	43	70	40
Soda	4	19	19	23	24	34	47	27
Yeast	13	62	25	30	4	6	42	24
Cream of Tartar + Soda	1	5	1	1	11	16	13	7.5
Soda + Yeast	0	—	0	—	1	1	1	.5
Total	21		82		70		173	
Total Non-Yeast	8	38	57	70	66	94	131	76

Source: Gillette, *White House Cook Book*, 227, 231–36.

Table A-6. Muffin Comparison Chart, 1920s/1975/1997

Date	Type	Flour (Cups)	Baking Powder (tsps.)	Sugar (Cups)	Eggs	Milk (Cups)	Butter/Oil (Tbls.)	Yield
1902	Standard[a]	1½	1	—	2	milk 1	—	1 doz.
1920s	Lose Weight	Graham 2	4	—	—	skim ¾–1	2	—
1920s	Gain Weight	2	4	¼	2	1	4	—
1975	Standard	1¾	2	¼	2	¾	2–4	2 doz.
1997	Low Fat	2	4	½–⅔	1	nonfat 1	3½	1 doz.
1997	Standard	2	3	⅔	2	cream 1	4–8	1 doz.

All recipes contained approximately the same amount of salt, which has been omitted here.

a. From Sarah Tyson Rorer, *Mrs. Rorer's New Cook Book* (Philadelphia: Arnold and Company, 1902), 511.

Notes

Introduction

1. William Rubel, *Bread: A Global History*, Edible Series, Andrew F. Smith, ed. (London: Reaktion Books, 2011), 10.

2. Ralph Waldo Emerson, "History," *Self-Reliance and Other Essays* (New York: Dover Publications, 1993), 1.

3. Carolyn Korsmeyer, *Making Sense of Taste: Food & Philosophy* (Ithaca, NY: Cornell University Press, 1999), senses, 84; art, 156–57.

4. Alfred D. Chandler Jr., *The Visible Hand: The Managerial Revolution in American Business* (Cambridge: Belknap Press, 1977), appendix A, 503–512.

5. Sidney Mintz, *Tasting Food, Tasting Freedom: Excursions into Eating, Culture, and the Past* (Boston: Beacon Press, 1996), 3.

6. Ann Romines, "Reading the Cakes: *Delta Wedding* and the Texts of Southern Women's Culture," *Mississippi Quarterly* 50, no. 4 (1997): 601.

Chapter 1. The Burden of Bread

1. Sarah Josepha Hale, *Early American Cookery: The "Good Housekeeper"* (1841; Mineola, NY: Dover Publications, 1996), 24–26.

2. Eliza Leslie, *Miss Leslie's New Cookery Book* (Philadelphia: T. B. Peterson and Brothers, 1857), 455, 467–68.

3. Hale, *Early American Cookery*, 121, 116.

4. Definitions from *Oxford English Dictionary*, 1:1557, 1663.

5. The Assize of Bread weight definitions for loaves were in effect until 2008. Harry Wallop, "Bread Rules Abandoned after 750 Years," *Telegraph*, Sept. 25, 2008.

6. Thomas Cogan, *The Haven of Health* (London: Printed by Anne Griffin, for Roger Ball, 1636), wheat, 27; rye, 29, https://archive.org/stream/havenofhealthchi00coga#page/26/mode/2up.

7. Ibid., 25.

8. William Rubel, *Bread: A Global History*, Edible Series, edited by Andrew F. Smith (London: Reaktion Books, 2011), baguette, 128; yeast, 17–18.

9. M. L. Pasteur, *Mémoire sur la Fermentation Alcoolique* (Paris: Imprimerie de Mallet-Bachelier, 1860), http://archive.org/stream/mmoiresurlafer00past#page/n3/mode/2up.

10. Eliza Leslie, *Seventy-Five Receipts for Pastry, Cakes and Sweetmeats*, 3rd ed. (Boston: Munroe and Francis, 1828), 63.

11. Ibid., 62.

12. Anon., *The Cook Not Mad* (Watertown, NY: Knowlton & Rice, 1831), 80.

13. Lydia Marie Child, *The American Frugal Housewife*, 2nd ed. (1829; Mineola, NY: Dover Publications, 1999), 11, http://digital.lib.msu.edu/projects/cookbooks/books/frugalhouse wifechild/frch.pdf.

14. Catharine Beecher, *Miss Beecher's Domestic Receipt-Book* (1841; Mineola, NY: Dover, 2001), 85–86, 147.

15. Ibid., 230.

16. Amelia Simmons, *The First American Cookbook: A Facsimile of "American Cookery"* (1796; New York: Dover Publications, 1958), 64.

17. William Cronon, *Changes in the Land: Indians, Colonists, and the Ecology of New England*. (New York: Hill and Wang, 1983), 154.

18. James E. McWilliams, *A Revolution in Eating: How the Quest for Food Shaped America* (New York: Columbia University Press, 2005), 82.

19. Ibid., 55–56.

20. Ibid., 15.

21. Ibid., 83.

22. Maria Eliza Ketelby Rundell, *A New System of Domestic Cookery* (Philadelphia: Benjamin C. Buzby, 1807), 243, http://digital.lib.msu.edu/projects/cookbooks/books/newsystem domestic/syst.html#syst221.jpg.

23. Karen Hess, transcr., *Martha Washington's Booke of Cookery and Booke of Sweetmeats* (New York: Columbia University Press, 1981), 447–63.

24. Martha Ballard's Diary Online, http://dohistory.org/diary/index.html.

25. Stephen Nissenbaum, *Sex, Diet, and Debility in Jacksonian America: Sylvester Graham and Health Reform* (Chicago: Dorsey Press, 1980), 3.

26. Gamber overlooks these in her book *The Boardinghouse in Nineteenth-Century America* (Baltimore: Johns Hopkins University Press, 2007).

27. William Andrus Alcott, *The Young House-Keeper or Thoughts on Food and Cookery*, 6th ed. (Boston: Waite Peirce, and Co., 1846), http://digital.lib.msu.edu/projects/cookbooks/books/younghousekeeper/youn.html#youn101.gif.

28. Ibid., 105–106.

29. Edward Hitchcock, *Dyspepsy Forestalled and Resisted: or Lectures on Diet, Regimen, and Employment; delivered to the students of Amherst College, Spring Term, 1830* (Amherst, MA: J. S. & C. Adams, 1831), 27–28.

30. Ibid., 34.

31. Ibid., 67.

32. Mme. Mérigot, *La Cuisiniere Républicaine, Qui enseigne la manière simple d'accommoder les Pommes de terre* (Paris: Chez Mérigot jeune, libraire, L'an III de la République [1794 or 1795]), in *Les Liaisons Savoureuses: Réflexions et pratiques culinaires au XVIIIe Siècle*, edited by Beatrice Fink, 169–85 (Saint-Étienne, France: Publications de l'Université de Saint-Étienne, 1995).

33. Andrew Dalby, *Flavours of Byzantium* (Devon, England: Prospect Books, 2003), has specific foods and humoral theory, and a month-by-month calendar of food and physical activity. The first cookbook printed on a printing press, in Florence, Italy, ca. 1474, is also an upper-class male manual, Platina's *De honesta voluptate et valetudine* (On Right Pleasure and Good Health), in *The Art of Cooking: The First Modern Cookery Book*, edited by Luigi Ballerini, translated by Jeremy Parzen (Berkeley: University of California Press, 2005).

34. Nancy Cott, *Bonds of Womanhood: "Woman's Sphere" in New England, 1780–1835* (New Haven, CT: Yale University Press, 1977), 101. In her groundbreaking discussion of how the philosophy of domesticity was transmitted to women, Cott does not include cookbooks or books of household management.

35. Susan Strasser, *Never Done: A History of American Housework* (New York: Pantheon Books, 1982).

36. Ruth Schwartz Cowan, *More Work for Mother: The Ironies of Household Technology from the Open Hearth to the Microwave* (New York: Basic Books, 1983).

37. Glenna Matthews, *"Just a Housewife": The Rise and Fall of Domesticity in America* (New York: Oxford University Press, 1987).

38. Janet Theophano, *Eat My Words: Reading Women's Lives through the Cookbooks They Wrote* (New York: Palgrave, 2002).

39. Natalie Zemon Davis, *Women on the Margins: Three Seventeenth-Century Lives* (Cambridge, MA: Harvard University Press, 1995).

40. Simmons, *"American Cookery."*

41. Mary Randolph, *The Virginia Housewife; or, Methodical Cook* (1831; New York: Dover, 1993).

42. Eliza Leslie, *Miss Leslie's New Cookery Book* (Philadelphia: T. B. Peterson and Brothers, 1857).

43. Child, *American Frugal Housewife*.

44. Beecher, *Miss Beecher's Domestic Receipt-Book*.

45. Anon., *Cook Not Mad*, iv; italics in original.

46. Ibid., iii.

47. Martha Ballard, Martha Ballard's Diary Online, entry for Wed., May 27, 1807.

48. Simmons, *"American Cookery,"* 44.

49. Ruth E. Finley, *The Lady of Godey's: Sarah Josepha Hale* (Philadelphia: J. B. Lippincott, 1931), 68.

50. Hale, *Early American Cookery*, 27.

51. Child, *American Frugal Housewife*, http://www.gutenberg.org/files/13493/13493-h/13493-h.htm, 5/30/2013.

52. Hale, *Early American Cookery*, 95.

53. Ibid., 31, 22.

54. Janice (Jan) Bluestein Longone, introduction to Beecher, *Miss Beecher's Domestic Receipt-Book*, iv.

55. Kathryn Kish Sklar, *Catharine Beecher: A Study in American Domesticity* (New York: W. W. Norton & Co., 1976), 113.

56. Mark McWilliams, "Good Women Bake Good Biscuits: Cookery and Identity in Antebellum American Fiction," *Food, Culture & Society* 10, no. 3 (2007): 388–406.

57. Margaret Beetham, "Of Recipe Books and Reading in the Nineteenth Century: Mrs. Beeton and her Cultural Consequences," in *The Recipe Reader: Narratives, Contexts, Traditions*, edited by Janet Floyd and Laurel Forster (Lincoln: University of Nebraska Press, 2003), 20.

58. Harriet Beecher Stowe, *The Minister's Wooing* (New York: Penguin, 1859), 12.

59. Gamber, *Boardinghouse*, 82–83.

60. Kim Cohen, "'True and Faithful in Everything': Recipes for Servant and Class Reform in Catherine Owen's Cookbook Novels," in *Culinary Aesthetics and Practices in Nineteenth-Century American Literature*, edited by Monika Elbert and Marie Drews (New York: Palgrave Macmillan, 2009), 107.

61. William G. Panschar, *Baking in America: Economic Development*, vol. 1 (Evanston, IL: Northwestern University Press, 1956), 35.

62. Ibid., 30–31.

63. Ibid., 34.

64. "Wetting" is liquid such as water or milk.

65. Beecher, *Domestic Receipt-Book*, 87.

66. Mary Bellis, "The History of the Thermometer," About Money, http://inventors.about.com/od/tstartinventions/a/History-Of-The-Thermometer.htm.

67. Leslie, *New Cookery Book*, 435.

68. Beecher, *Domestic Receipt-Book*, 83–84.

69. Hess, *Martha Washington's Booke of Cookery*, 316–17.

70. Leslie, *New Cookery Book*, 1857, 523.

71. Hale, *Early American Cookery*, 96. Most raisins were from Muscat of Alexandria grapes with seeds until the 1920s, when Thompson Seedless grapes replaced them. L. Peter Christensen, "Raisin Grape Varieties," http://iv.ucdavis.edu/files/24430.pdf.

72. Hale, *Early American Cookery*, 516.

73. Hannah Glasse, *The Art of Cookery Made Plain and Easy* (1747; Bedford, MA: Applewood Books, 1997), 162. Also Hale, *Early American Cookery*, 101.

74. Leslie, *New Cookery Book*, 517.

Chapter 2. The Liberation of Cake

1. Harry Miller, "Potash from Wood Ashes: Frontier Technology in Canada and the United States," *Technology and Culture* 21, no. 2 (1980): 198. The chemical formula for potassium carbonate is K_2CO_3.

2. Hannah Glasse, *The Art of Cookery Made Plain and Easy*, historical notes by Karen Hess (1747; Bedford, MA: Applewood Books, 1997), x.

3. Donna R. Barnes and Peter G. Rose, *Matters of Taste: Food and Drink in Seventeenth-Century Dutch Art and Life* (Albany, NY: Albany Institute of History and Art, 2002), 34.

4. The First United States Patent Statute, Patent Act of 1790, ch. 7, 1 stat. 109 (April 10, 1790); http://docs.law.gwu.edu/facweb/claw/patact1790.htm.

5. "First U.S. Patent Issued Today in 1790," July 31, 2001, United States Patent and Trademark Office, Press Release #01–33; http://www.uspto.gov/news/pr/2001/01-33.jsp.

6. Waverly Root and Richard de Rochemont, *Eating in America* (Hopewell, NJ: Ecco Press, 1995), 138.

7. Miller, "Potash from Wood Ashes," 198n23.

8. Lydia Maria Child, *The American Frugal Housewife* (1829; Mineola, NY: Dover Publications, 1999), 29.

9. Amelia Simmons, *American Cookery*, 2nd ed. (Hartford, CT: Printed for Simeon Butler, Northampton, 1798), 49–50, http://digital.lib.msu.edu/projects/cookbooks/html/books/book_01.cfm.

10. Ibid.

11. Mary Tolford Wilson, "The First American Cookbook," in Simmons, *American Cookery*, 1st ed., (1796; New York: Dover Publications, 1958), xiv–xv.

12. Anon., "No 125. Muffins," *The Cook Not Mad* (Watertown, NY: Knowlton & Rice, 1831).

13. Sarah Josepha Hale, *Early American Cookery: The "Good Housekeeper"* (1841; Mineola, NY: Dover Publications, 1996), 24–27.

14. Like British plum pudding, plum cakes contained no plums; the word was used in the sense of "rich, good, desirable." *Oxford English Dictionary.*

15. Hale, *Early American Cookery*, 99.

16. Campbell Gibson, *American Demographic History Chartbook, 1790–2010*, http://www.demographicchartbook.com/Chartbook/images/chapters/gibson06.pdf.

17. From twenty days to six. David M. Kennedy, Lizabeth Cohen, Thomas A. Bailey, *The American Pageant: A History of the Republic*, 12th ed., vol. 1 (New York: Houghton Mifflin, 2002), 312. Rochester wheat is discussed in Paul E. Johnson, *A Shopkeeper's Millennium* (1978; New York: Hill and Wang, 2004).

18. Johnson, *Shopkeeper's Millennium*, 18.

19. Catharine Beecher, *Miss Beecher's Domestic Receipt-Book* (1841; Mineola, NY: Dover Publications, 2001), 229.

20. Eliza Leslie, *Miss Leslie's New Cookery Book* (Philadelphia: T. B. Peterson and Brothers, 1857), 433.

21. Martha Ballard, Martha Ballard's Diary Online, 1785–1812, http://dohistory.org/diary; Mary Randolph, *The Virginia Housewife; or, Methodical Cook* (1831; New York: Dover Publications, 1993), 135.

22. Ballard, Martha Ballard's Diary Online; Sandra L. Oliver, *Saltwater Foodways: New Englanders and Their Food, at Sea and Ashore, in the Nineteenth Century* (Mystic, CT: Mystic Seaport Museum, 1995), 106–108.

23. Charles Henry Dana, *Two Years Before the Mast*, 1840. https://en.wikisource.org/wiki/Two_Years_Before_the_Mast/Chapter_V.

24. *Martha Washington's Booke of Cookery and Booke of Sweetmeats*, transcribed by Karen Hess (New York: Columbia University Press, 1981), 337.

25. Randolph, *Virginia Housewife*, 139–40.

26. Leslie, *New Cookery Book*, 432.

27. Ibid., 426–27.

28. Minnie C. Fox, comp., *The Blue Grass Cook Book*, new introduction by Toni Tipton-Martin (1904; Lexington: University Press of Kentucky, 2005), xviii–xxi.

29. Jean Fagan Yellin, *Women & Sisters: The Antislavery Feminists in American Culture* (New Haven, CT: Yale University Press, 1989), 53.

30. "Spirit of Hartshorn," Dictionary.com, http://dictionary.reference.com/browse/spirit+of+hartshorn. The chemical formula for ammonia gas is NH_3.

31. The chemical formula for ammonium carbonate is $(NH_4)_2CO_3$; "Ammonium Carbonate," The Chemical Company; http://www.thechemco.com/chemical/ammonium-carbonate.

32. "Ammonium Carbonate," English Oxford Living Dictionaries; http://oxforddictionaries.com/definition/english/ammonium%2Bcarbonate.

33. Eben N. Horsford, *The Theory and Art of Bread-Making: A New Process without the Use of Ferment* (Cambridge, MA: Welch, Bigelow, and Co., Printers to the University, 1861), 161.

34. Email from Kerry Group to author, Nov. 14, 2012. The only Borwick historic records in existence are two pages handwritten on April 16, 1943. This was an attempt to memorialize what was known about the founding of the company, because the "premises were entirely destroyed" in the German Luftwaffe Blitz attack on London on December 29, 1940.

35. Paul R. Jones, "Justus Von Liebig, Eben Horsford, and the Development of the Baking Powder Industry," *Ambix* 40, part 2 (July 1993): 70.

36. Alfred D. Chandler Jr., "The Enduring Logic of Industrial Success," *Harvard Business Review*, March–April 1990, 133.

37. The chemical formula for sodium bicarbonate is $CHNaO_3$; "Sodium Bicarbonate," NIH U.S. National Library of Medicine, National Center for Biotechnology Information, Pub Chem Open Chemistry Database, http://pubchem.ncbi.nlm.nih.gov/summary/summary.cgi?cid=516892; "About Us," Arm & Hammer; http://armandhammer.ca/about-us.

38. The chemical formula for potassium bicarbonate is $CHKO_3$; "Potassium Bicarbonate," NIH U.S. National Library of Medicine.

39. "Quick Breads," Food Reference.com, http://www.foodreference.com/html/fquickbreads.html.

40. Leslie, *New Cookery Book*, 407–434. Leslie adds this tip at the end of most bread recipes.

41. The chemical formula for ammonium carbonate is Na_2CO_3; "Ammonium Carbonate," English Oxford Living Dictionaries, http://oxforddictionaries.com/definition/english/ammonium%2Bcarbonate.

42. Beecher, *Domestic Receipt-Book*, 202.

43. Ibid., 89.

44. Elizabeth M. Hall, comp., *Practical American Cookery and Domestic Economy* (New York: C. M. Saxton, Barker & Co., 1860), 65.

45. Ibid., 273.

46. Leslie, *New Cookery Book*, 433.

47. By the early twentieth century, "ammonia ha[d] been abandoned." A. C. Morrison Testimony before U.S. Industrial Commission, May 17, 1901, Washington, DC, in A. Cressy Morrison, *Baking Powder Controversy* (hereafter, BPC), 210–11.

48. "Ammonium Carbonate," The Chemical Company; http://www.thechemco.com/chemical/ammonium-carbonate.

49. Hale, *Early American Cookery*, 110; italics in original.

50. Child, *American Frugal Housewife*, brandy, 18; rum, 74; beer, 86.

51. Beecher, *Domestic Receipt-Book*, 202.

52. Leslie, *New Cookery Book*, 425.

53. Beecher, *Domestic Receipt-Book*, 131.

54. Anon., *Cook Not Mad*, 75–76.

55. Beecher, *Domestic Receipt-Book*, dampness, 233; stomach, 137; italics in original.

56. Ibid., 228. The alkali baking soda is routinely used now to soothe stomach problems.

57. Child, *American Frugal Housewife*, 15.

58. Eliza Leslie, *Seventy-Five Receipts for Pastry, Cakes, and Sweetmeats*, 3rd ed. (Boston: Munroe and Francis, 1828), 66.

59. Ibid., 67.

60. Ibid., 70, 68.

61. Stephen Schmidt, "Cakes," in Andrew F. Smith, ed., *The Oxford Encyclopedia of Food and Drink in America* (New York: Oxford University Press, 2004), 157–58.

62. Child, *American Frugal Housewife*, 71.

63. Harriet Beecher Stowe, *The Minister's Wooing* (New York: Penguin, 1859), 11.

64. Martha Ballard, Martha Ballard's Diary Online, 1785–1812.

65. Schmidt, "Cakes," 158.

66. Beecher, *Domestic Receipt-Book*, 241.

67. Hale, *Early American Cookery*, 96.

68. Ibid.

69. Ibid., 31.

70. Glasse, *Art of Cookery*, ix.

71. John L. Hess and Karen Hess, *The Taste of America* (Urbana: University of Illinois Press, 2000), 92.

72. Ibid., 57.

73. Harold McGee, *On Food and Cooking: The Science and Lore of the Kitchen* (New York: Scribner, 2004), 559.

74. Hess and Hess, *Taste of America*, 92.

75. Karen Hess, introduction to Simmons, *American Cookery*, 69.

76. Glasse, *Art of Cookery*, ix.

Chapter 3. The Rise of Baking Powder Business: The Northeast

1. The slogan is "Better things for better living through chemistry." Faye Rice, "Dow Chemical: From Napalm to Nice Guy," *Fortune*, May 12, 1986, http://money.cnn.com/magazines/fortune/fortune_archive/1986/05/12/67544/index.htm.

2. Steven Shapin, "The Medical Making of Modernity: Knowing about Our Food, Our Bodies, and Ourselves over the Past 2,000 Years," paper presented at the New York Academy of Sciences, Feb. 2, 2011, 3.

3. *Oxford English Dictionary*; Thomas Kuhn, quoted in Michael D. Dahnke and H. Michael Dreher, *Philosophy of Science for Nursing Practice: Concepts and Application* (New York: Springer Publishing, 2011), 87.

4. American Philosophical Museum, http://www.apsmuseum.org.

5. "The Chemical Revolution of Antoine-Laurent Lavoisier," American Chemical Society, http://www.acs.org/content/acs/en/education/whatischemistry/landmarks/lavoisier.html.

6. Bernard Jaffe, *Men of Science in America: The Role of Science in the Growth of Our Country* (New York: Simon and Schuster, 1946), 65.

7. Melody Kramer, "Meet 115, the Newest Element on the Periodic Table," National Geographic, August 30, 2013, http://news.nationalgeographic.com/news/2013/08/130828-science-chemistry-115-element-ununpentium-periodic-table.

8. "Antoine-Laurent Lavoisier," Chemical Heritage Foundation, Sept. 11, 2015, http://www.chemheritage.org/discover/online-resources/chemistry-in-history/themes/early-chemistry-and-gases/lavoisier.aspx.

9. Theodore M. Porter, *Trust in Numbers: The Pursuit of Objectivity in Science and Public Life* (Princeton, NJ: Princeton University Press, 1995), 14–15.

10. *Oxford English Dictionary*, vol. 1 (New York: Oxford University Press, 1981), 390.

11. Amelia Simmons, *The First American Cookbook: A Facsimile of "American Cookery,"* (1796; New York: Dover Publications, 1958), 6.

12. Frederick Accum, *A Treatise on Adulterations of Food and Culinary Poisons* (Philadelphia: Ab'm Small, 1820), http://www.gutenberg.org/files/19031/19031-h/19031-h.htm.

13. *Deadly Adulteration and Slow Poisoning Unmasked; or, Disease and Death in the Pot and the Bottle* (London: Sherwood, Gilbert, and Piper, ca. 1830), 70.

14. Ibid., 73.

15. Florence Nightingale, "Taking Food," in *Directions for Cooking by Troops in Camp and Hospital* (Richmond, VA: J. W. Randolph, 1861), 29–30.

16. Uschi Schling-Brodersen, "Liebig's Role," *Ambix* 39, part 1 (March 1992): 26.

17. "History," National Academy of Sciences, http://www.nasonline.org/about-nas/history.

18. Justus von Liebig, *Researches on the Chemistry of Food and the Motion of the Juices in the Animal Body* (Lowell, MA: Daniel Bixby and Co., 1848).

19. "Liebig's Dietetic Trinity," in ibid.; see Food Revolutions/Science and Nutrition, 1700–1950, http://library.missouri.edu/exhibits/food/liebig.html.

20. "Eben N. Horsford," Rensselaer Polytechnic Institute Alumni Hall of Fame, http://www.rpi.edu/about/alumni/inductees/horsford.html.

21. Samuel Rezneck, "The European Education of an American Chemist and Its Influence in 19th-Century America: Eben Norton Horsford," *Technology and Culture* 11, no. 3 (1970): 366–88.

22. Jaffe, *Men of Science*, 67.

23. Ibid., 72.

24. "Development of Baking Powder," American Chemical Society National Historic Chemical Landmarks, http://www.acs.org/content/acs/en/education/whatischemistry/landmarks/bakingpowder.html.

25. Clark A. Elliott and Margaret W. Rossiter, *Science at Harvard: Historical Perspectives* (Bethlehem, PA: Lehigh University Press 1992), 70.

26. "Horsford, Eben Norton," *Complete Dictionary of Scientific Biography*, Encyclopedia.com, http://www.encyclopedia.com/doc/1G2-2830902062.html.

27. Elliott and Rossiter, *Science at Harvard*,70.

28. Rhode Island Historical Society, Rumford Chemical Works Historical Note, Dec. 6, 1976, 1.

29. Daniel Walker Howe, *Making the American Self: Jonathan Edwards to Abraham Lincoln* (Cambridge, MA: Harvard University Press, 1997), 136.

30. Now it is called calcium acid phosphate.

31. On experimentation, see Paul R. Jones, "Justus von Liebig, Eben Horsford, and the Development of the Baking Powder Industry," *Ambix* 40, part 2 (July 1993): 67. E. N. Horsford, Patent 14,722, April 22, 1856, 1. "Improvement in Preparing Phosphoric Acid as a Substitute for Other Solid Acids," April 22, 1856, United States Patent and Trademark Office, http://

pdfpiw.uspto.gov/.piw?Docid=14722&idkey=NONE&homeurl= http%3A%252F%252Fpatft .uspto.gov%252Fnetahtml%252FPTO%252Fpatimg.htm.

32. Horsford Patent 14,722.

33. Rumford Chemical Works Archive (hereafter, RCW) series A, vol. A Stockholders and Directors' Records Book A, 1859–1883, April 1, 1859, 2.

34. William Rubel, *Bread: A Global History* (London: Reaktion Books, 2011), 73.

35. Ibid., 146.

36. Horsford, *The Theory and Art of Bread-Making: A New Process without the Use of Ferment* (Cambridge, MA: Welch, Bigelow, and Co., 1861), 17.

37. Isabella Beeton, *Mrs. Beeton's Book of Household Management* (London, 1861), http://www .mrsbeeton.com/35-chapter35.html.

38. Horsford, *Theory and Art of Bread-Making*, 11.

39. Ibid., 11.

40. Ibid., 7.

41. Ibid., 21–22.

42. Ibid., 26.

43. Ibid., 27.

44. *Wholesome Cooking with Horsford's Bread Preparation*, ca. 1920s, HC 63, Margaret Dotson Foodways Collection, Berea College Historical Collections; http://digital.berea.edu/cdm/ref/ collection/p16020coll1/id/1419.

45. "Development of Baking Powder," American Chemical Society.

46. RCW series A, vol. A Stockholders and Directors' Records Book A, 1859–1883, Nov. 6, 1865, n.p.

47. "About Us," History of Argo, http://www.argostarch.com/about_us.html.

48. Melita Marie Garza, "Unilever to Sell Argo Cornstarch," *Chicago Tribune*, June 1, 2001, http://articles.chicagotribune.com/2001-06-01/business/0106010222_1_unilever-brands -starch.

49. "Development of Baking Powder," American Chemical Society.

50. Ibid.

51. "Development and Use of Baking Powder and Baking Chemicals," United States Department of Agriculture, Circular No. 138 (Washington, DC, Nov., 1930), 3.

52. Jones, "Justus Von Liebig," 72.

53. "Notes," *Journal of the Society of Arts* 17, no. 879 (September 24, 1869): 840, https://www .jstor.org/stable/pdf/41334796.pdf.

54. Jones, "Justus Von Liebig," 73.

55. Ibid., 72.

56. Ibid., 73.

57. P. Christiaan Klieger, *Images of America: The Fleischmann Yeast Family* (Charleston, SC: Arcadia Publishing, 2004), 16.

58. Jones, "Justus Von Liebig," 71–72.

59. RCW series D, vol. 93, Baking Powder Experiments 1889–1895.

60. Stockholders' Meeting on November 6, 1876, RCW series A, vol. A, Stockholders and Directors' Records Book A, 1859–1883.

61. RCW series A, vol. A, Stockholders and Directors' Records Book A, 1859–1883.

62. Sarah Tyson Rorer, *Mrs. Rorer's New Cook Book: A Manual of Housekeeping* (Philadelphia: Arnold and Co., 1902), 507.

63. *Ziegler vs. Hoagland, et al.,* NY 1888, 606–607. A Mr. F. H. Hall was briefly involved in the company. There is a discrepancy about when Ziegler and the Hoaglands met. Joseph Hoagland testified that he met Ziegler within a day or two of arriving in New York City in October 1868 (A. Cressy Morrison, *The Baking Powder Controversy* [New York: American Baking Powder Association, 1904–1907; hereafter BPC], 607), Ziegler testified that he met the Hoaglands in 1870 (BPC, 595).

64. "History of Beaver County," Beaver County, PA, http://www.beavercountypa.gov/history-beaver-county. Different credentials appear in "William Ziegler Dead after a Long Illness," *Evening Telegram*, New York, May 24, 1905; BPC, 1951. This obituary does not show the pharmacy school but has Ziegler attending Eastman's business college in Poughkeepsie, New York.

65. *Ziegler vs. Hoagland, et al.,* 1888; BPC, 600.

66. BPC, 584.

67. Ibid., 608.

68. Sarah J. Hale and Louis A. Godey, eds. *Godey's Lady's Book and Magazine* 88, Jan.–June 1874 (Philadelphia: Louis A. Godey, 1874), 189–90.

69. "Martha Washington Tea Party," *Cambridge Chronicle* 31, no. 9 (Feb. 26, 1876), 1.

70. "Martha Washington Balls and Tea Parties," *Masquerades, Tableaux, and Drills* (New York: Butterick Publishing Co., 1906), 12, http://memory.loc.gov/cgi-bin/ampage?collId=musdi&fileName=033/musdi033.db&recNum=14.

71. Ibid.

72. "Martha Washington Tea Party," *Pittsburgh Sunday Press Illustrated Magazine Section*, Feb. 20, 1910, front page. In 2013 the Connecticut Daughters of the American Revolution held a Martha Washington Tea Party as a fund-raiser. Lisa Saunders, "Martha Washington Tea Party & Fundraiser," Stonington Patch, Sept. 25, 2013, http://stonington.patch.com/groups/events/p/martha-washington-tea-party; Washington's Birthday Celebration Association, "The Society of Martha Washington Cocktail Party," Feb. 18, 2017, Laredo, Texas, Country Club, http://www.wbcalaredo.org/event/the-society-of-martha-washington-cocktail-party.

73. Janice Bluestein (Jan) Longone, "Cookbooks and Manuscripts: From the Beginnings to 1860," in *The Oxford Encyclopedia of Food and Drink in America*, vol. 1., edited by Andrew F. Smith (New York: Oxford University Press, 2004), 291.

74. Elizabeth Duane Gillespie, comp., *National Cookery Book, compiled from original receipts, for the Women's Centennial Committees of the International Exposition.* Philadelphia: Women's Centennial Executive Committee. Introduction by Andrew F. Smith (Bedford, MA: Applewood Books, n.d. [1876]), iv.

75. Ibid., 163–82, 209–225, 240–41, 255–68.

76. Gillespie, *National Cookery Book*, "Feather Cake," 181; "Hocus Pocus," 255. Azumea was created by Professor William Morris. The brand had its own cookbook, *Professor Morris' Azumea, the Premium Baking Powder* (Philadelphia: E.W.P. Taunton, 1867).

77. Women of the First Congregational Church, Marysville, Ohio, *Centennial Buckeye Cook Book*(1876; Columbus: Ohio State University Press, 2000), ix.

78. Smith, *Oxford Encyclopedia of Food and Drink*.

79. Ibid., 23–27.

80. Catharine Beecher, *Miss Beecher's Domestic Receipt-Book* (1858; Mineola, NY: Dover, 2001), 146; Women of the First Congregational Church, *Centennial Buckeye Cook Book,* 56.

81. Women of the First Congregational Church, *Centennial Buckeye Cook Book,* xxv.

82. "Table A. Mean Center of Population of the United States: 1790 to 1990," U.S. Census, https://www.census.gov/population/www/censusdata/files/popctr.pdf.

Chapter 4. The Advertising War Begins

1. James D. Norris, *Advertising and the Transformation of American Society, 1865–1920* (New York: Greenwood Press, 1990), xvii.

2. Harvey A. Levenstein, *A Revolution at the Table: The Transformation of the American Diet* (New York: Oxford University Press, 1988), 32.

3. James Harvey Young, *The Toadstool Millionaires: A Social History of Patent Medicines in America before Federal Regulation* (Princeton, NJ: Princeton University Press, 1961), 68.

4. Lorine Swainston Goodwin, *The Pure Food, Drink, and Drug Crusaders, 1879–1914* (Jefferson, NC: McFarland, 1999), 65.

5. Young, *Toadstool Millionaires*, 105.

6. *Ziegler v. Hoagland, et al., NY* 1888; BPC, 602.

7. Ibid., 609.

8. "The Baking Powder War," *Daily Eagle*, Dec. 21, 1878, 3.

9. 1876 or 1877, Ziegler was not sure which year; *Ziegler v. Hoagland*.

10. *Ziegler v. Hoagland*, May 28, 1888; BPC, 609, 627.

11. Joseph M. Gabriel, "Restricting the Sale of 'Deadly Poisons': Pharmacists, Drug Regulation, and Narratives of suffering in the Gilded Age," *Journal of the Gilded Age and the Progressive Era* 9, no. 30 (2010): 313–36.

12. "Baking Powder War," *Daily Eagle*.

13. "Baking Powders in the United States," *New York Tribune*, in *The Analyst*, edited by G. W. Wigner and J. Muter, Royal Society of Chemistry (London: Baillière, Tindall, and Cox, 1881), 91–93; "Condemned: Alum Baking Powders in Court—Interesting Testimony of Scientific Men," *Sacramento Daily Union* 13, no. 13, March 8, 1881, California Digital Newspaper Collection, http://cdnc.ucr.edu/cgi- bin/cdnc?a=d&d=SDU18810308.2.22.

14. *Ziegler v. Hoagland*; Hoagland claim, BPC, 609–10; Ziegler claim, 627–28.

15. Giuseppi Rudmani, comp., *The Royal Baker and Pastry Cook* (New York: Royal Baking Powder Company, 1878).

16. Elizabeth Duane Gillespie, comp., *National Cookery Book* (1876; Bedford, MA: Applewood Books, 2010), 163.

17. Andrew F. Smith, "Advertising and Promotional Cookbooklets in the Nineteenth Century," paper presented at the Cookbook Conference, New York, Feb. 9, 2012, 1–2.

18. Young, *Toadstool Millionaires*, 140–41.

19. Rudmani, *Royal Baker and Pastry Cook*, 2 pages after cover.

20. Ibid., 1 page after cover.

21. "The Real Isabella Beeton: A Biography with Recipes," *The Secret Life of Mrs. Beeton*, PBS, http://www.pbs.org/wgbh/masterpiece/mrsbeeton/beeton.html.

22. Rudmani, *Royal Baker and Pastry Cook*, 8, 9, 27, 23.

23. Ibid., 3 pages after cover.

24. Ibid.

25. "Because You're Worth It: The Story behind the Legendary Phrase," About L'Oréal Paris, http://www.lorealparisusa.com/en/About-Loreal-Paris/Because-Youre-Worth-It.aspx. The phrase was written in 1973 by Ilon Specht, a twenty-three-year-old female copywriter at McCann Erickson, for their client L'Oréal.

26. Rudmani, *Royal Baker and Pastry Cook,* 2 pages after cover.

27. Kathleen Brown, "The Maternal Physician: Teaching American Mothers to Put the Baby in the Bathwater," in *Right Living: An Anglo-American Tradition of Self-Help Medicine and Hygiene,* edited by Charles E. Rosenberg (Baltimore: Johns Hopkins University Press, 1993), 92–93.

28. Henry A. Mott Jr., PhD, EM, "The Deleterious Use of Alum in Bread and Baking Powders—Alum Being Substituted for Cream of Tartar," *Scientific American,* Nov. 16, 1878, 308, http://archive.org/stream/scientific-american-1878-11-16/scientific-american-v39-n20-1878-11-16_djvu.txt.

29. Mitchell Okun, *Fair Play in the Marketplace: The First Battle for Pure Food and Drugs* (DeKalb: Northern Illinois University Press, 1986), 235.

30. Ellen H. Richards, *The Chemistry of Cooking and Cleaning: A Manual for Housekeepers* (Boston: Estes & Lauriat, 1882); quoted in Claudia Quigley Murphy, *A Collation of Cakes Yesterday and Today, in Which Is Included a True and Accurate Notation of Early English and Colonial America Receipts, Showing the Beginning and Progress of the Gentle Art of Cake Making to the Present Time* (New York, 1923), 20.

31. Ellen Gruber Garvey, *The Adman in the Parlor: Magazines and the Gendering of Consumer Culture, 1880s to 1910s* (New York: Oxford University Press, 1996), 19–20.

32. Robert Jay, *The Trade Card in Nineteenth-Century America (*Columbia: University of Missouri Press, 1987), 3.

33. "Currier & Ives—The History of the Firm," The History of Currier & Ives, http://www.currierandives.com.

34. Garvey, *Adman in the Parlor,* 16–17.

35. Jay, *Trade Card,* 61.

36. Letter dated March 21, 1887, from N. D. Arnold to Josselyn and Handke, Rumford Chemical Works Archive (hereafter, RCW), series E, box 4.

37. http://www.currierandives.com.

38. These cards are all undated. Unless stated otherwise, all trade cards discussed and depicted in this section are from the Nahum (Nach) Waxman Collection of Food and Culinary Trade Cards, Division of Rare and Manuscript Collections, Cornell University Library.

39. RCW, series A, vol. B, 1883–1906, n.p.

40. RCW, series A, vol. B, 1883–1906, Feb. 6, 1884, 16.

41. Young, *Toadstool Millionaires,* 143.

42. Gabriel, "Restricting the Sale," 316.

43. Mark Pendergrast, *For God, Country, and Coca-Cola: The Definitive History of the Great American Soft Drink and the Company That Makes It* (New York: Basic Books, 2000), 14–15.

44. Lily Haxworth Wallace, *The Rumford Complete Cook Book* (1908; Providence, RI: Rumford Chemical Works, 2000). The 2000 facsimile edition came with a caveat: "We only recommend the use of baking powder for cooking" (241).

45. Library: RCW series A, vol. A, 1859–1883, Feb. 1, 1882, n.p; cemetery: RCW series A, vol. B, 1883–1906, Feb. 15, 1884, 17; hospital: RCW series A, vol. B, Stockholders' and Directors' Record Book B, 1883–1906, Dec. 22, 1885, 41.

46. RCW series A, vol. B, Stockholders' and Directors' Record Book B, 1883–1906, March 19, 1886, 44.

47. "In Memoriam, Eben Norton Horsford," Wellesley College, 1893, 31–32. https://archive.org/details/inmemoriamebennoo0well, Nov. 2, 2013.

48. Wellesley College Archives, Records of the Class of 1886, 1882–1953: A guide, Wellesley College website, http://academics.wellesley.edu/lts/archives/6c_classes/6C.1886.html.

49. "CONDEMNED. Alum Baking Powders in Court—Interesting Testimony of Scientific Men," *[Brooklyn] Daily Eagle,* Dec. 9, 1880, 1; reprt. in *New York Times*, Dec. 11, 1880, 10; *Salt Lake Herald,* Salt Lake City, UT, Dec. 24, 1880, 3; *Independent Record,* Helena, MT, Dec. 31, 1880, 3; *Black Hills Weekly Pioneer,* Deadwood, SD, Jan. 1, 1881, 1; *Gleason's Monthly Companion*, Boston, Jan. 1881, 188; *Los Angeles Herald* 14, no. 149, Jan. 23, 1881, 3; and others.

50. Dr. Price, *Table and Kitchen: A Compilation of Approved Cooking Receipts* (Chicago: Price Baking Powder Co., 1908).

51. "A Day at a Reservation: The Indian as He Is at Ross Fork Agency," *New York Times*, Oct. 1, 1889, 5.

52. Rev. Addison P. Foster, "The Dakota Indians," *American Missionary* 38, no. 6 (1884), 175.

53. Phyllis Hughes, ed., *Pueblo Indian Cookbook* (Santa Fe: University of New Mexico Press, 1982), 11.

54. "Piki Bread," Ark of Taste, Slow Food USA, http://www.slowfoodusa.org/ark-item/piki-bread.

55. Alice Ross, "Fry Bread," in Andrew F. Smith, ed., *The Oxford Encyclopedia of Food and Drink in America*, vol. 2. (New York: Oxford University Press, 2004), 169.

56. Tantri Wija, "Out of the Frying Pan," *Free New Mexican*, Aug. 16, 2006, 2. http://cretscmhd.psych.ucla.edu/healthfair/PDF%20articles%20for%20fact%20sheet%20linking/FryBread_health_AIAN.pdf.

57. *The Swedish-American Cookbook: A Charming Collection of Traditional Recipes Presented in Both Swedish and English* (1882; New York: Skyhorse Publishing, 2012).

58. *Ziegler v. Hoagland*; BPC, 609.

59. BPC, 616.

60. Mrs. F. L. Gillette, *White House Cook Book* (Chicago: R. S. Peale & Co., 1887), 232.

61. Ibid., 264.

Chapter 5. The Cream of Tartar Wars

1. U.S. Congress, Senate, Committee on Agriculture and Forestry, Pure Food Legislation: "Memorial of the American Baking Powder Association in the Matter of the Bill S. 3618," 56th Congress, 1st Session, Doc. No. 303, April 19, 1900, 63–64.

2. Fannie Merritt Farmer, *Original 1896 Boston Cooking-School Cookbook* (1896; Boston: Dover Publications, 1997).

3. Joel Shrock, *The Gilded Age* (Westport, CT: Greenwood Press, 2004), 103.

4. "Memorial of the American Baking Powder Association," 63–64.

5. Richard Tedlow, *New and Improved: The Story of Mass Marketing in America* (Boston: Harvard Business School Press, 1996), 16.

6. *Ziegler v. Hoagland*, BPC, 1, sales/profits, 612; advertising, 610–11.

7. Ibid., 603.

8. Ibid., 590.

9. Ibid., 591.

10. Ibid., 584, 587.

11. Ibid., 583.

12. "That Baking Powder Suit," *New York Times*, May 29, 1888, 9.

13. "Salaries Cut Down by the Court," *New York Times*, June 2, 1888, 8.

14. "William Ziegler Dies," *New York Times*, May 25, 1905, 9.

15. Sarah Tyson Rorer, *Mrs. Rorer's New Cook Book: A Manual of Housekeeping* (Philadelphia: Arnold and Co., 1902), 507.

16. "Just Now," *Brooklyn Eagle*, Feb. 21, 1897, 25.

17. "Advertisements, Circulars, Etc., Issued by Various Baking Powder Companies; Circular Published by Cleveland Baking Powder Company before the Combine," BPC, 2, 1621–23.

18. RCW, series A, vol. B, 1883–1906, Oct. 22, 1892, 109.

19. Patricia Shillingburg, "The Hoagland Brothers on Shelter Island from 1881 to 1896," Shelter Island Organization, 2003, http://www.shelter-island.org/hoagland_brothers.

20. Naomi R. Lamoreaux, *The Great Merger Movement in American Business, 1895–1904* (Cambridge, UK: Cambridge University Press, 1985), 1–2.

21. U.S. Congress, *Congressional Serial Set* (U.S. Government Printing Office, 1901), BPC, lxxxii.

22. Mary A. Yeager, *Competition and Regulation: The Development of Oligopoly in the Meat Packing Industry* (Greenwich, CT: JAI Press, 1981), 27.

23. Tedlow, *New and Improved*, 120.

24. "Baking Powder Combination," *Wall Street Journal*, March 3, 1899.

25. BPC, lxxxvii.

26. Lamoreaux, *Great Merger Movement*, 161.

27. Harvey A. Levenstein, *Fear of Food: A History of Why We Worry about What We Eat* (Chicago: University of Chicago Press, 2012), 47.

28. "Why Women Are Nervous," *British Medical News*, 1899, BPC, 1665.

29. Deborah Cadbury, *Chocolate Wars: The 150-Year Rivalry between the World's Greatest Chocolate Makers* (New York: Public Affairs, 2010), 135–37.

30. Whiskey includes bourbon, American distilled liquor made from corn. Whiskey is made from rye.

31. James Harvey Young, *Pure Food: Securing the Federal Food and Drugs Act of 1906* (Princeton, NJ: Princeton University Press, 1989), 165–68.

32. Ibid., 87–90.

33. Terri Lonier, "Alchemy in Eden: Entrepreneurialism, Branding, and Food Marketing in the United States, 1880–1920," *Enterprise & Society* 11, no. 4 (2010): 704.

34. Carolyn de la Peña, *Empty Pleasures: The Story of Artificial Sweeteners from Saccharin to Splenda* (Chapel Hill: University of North Carolina Press, 2010), 24.

35. "Paper on the Subject of Baking Powders as Affected by Proposed Pure Food Legislation Now Pending before Congress," "American Baking Powder Association in the Matter of the Bill S 3618," U.S. Senate, Committee on Agriculture and Forestry, Pure Food Legislation, April 21, 1900, 56th Congress, 1st Session, Washington, DC, BPC, 63.

36. *NIRA Hearing on Code of Fair Practices and Competition Presented by the Baking Powder Industry, held on April 30, 1934* (Washington, DC: J. L. Ward of Ward & Paul, official reporter, 1934), 5; italics in the original.

37. "Advertisements, Circulars, Etc.," BPC 2, 1626–27.

38. "Memorial of the American Baking Powder Association," quoting the *Syracuse Post-Standard*, March 28, 1900, BPC, 69.

39. Ibid., 72.

40. Ibid.

41. "Housekeeping Pests" and "Worse Than Sneak Thieves," in "Advertisements, Circulars, Etc.," BPC, 1624, 1626.

42. Ibid., "Burn the Samples," BPC, 1628.

43. Ibid., 1651–52.

44. Ibid., "Law of Wisconsin," 1639–43.

45. Introduction, BPC, 7.

46. A. Cressy Morrison, Address—"The Baking Powder Controversy," *Journal of Proceedings, Eighth Annual Convention and International Pure Food Congress of the National Association of State Dairy and Food Departments*, St. Louis, MO, Sept. 29,1904, BPC, 1189–90.

47. Andrew Warnes, "'Talking' Recipes: What Mrs. Fisher Knows and the African-American Cookbook Tradition," in *The Recipe Reader: Narratives, Contexts, Traditions*, ed. Janet Floyd and Laurel Forster (2003; Lincoln: University of Nebraska Press, 2010), 64–65.

48. Abby Fisher, *What Mrs. Fisher Knows about Old Southern Cooking* (1881; Bedford, MA: Applewood Books, 1995), 85; italics in original.

49. Letter from Geo. Wilson to Church & Clark, June 5, 1871, RCW.

50. Letter from Geo. Wilson to Church & Clark, March 21, 1873, RCW.

51. United States Patent Office, Alexander P. Ashbourne, of Oakland, California, "Improvement in Biscuit Cutters," May 11, 1875. I discovered Ashbourne in Ida B. Wells, ed., "The Reason Why the Colored American Is Not in the World's Columbian Exposition," 1893, http://www.digital.library.upenn.edu/women/wells/exposition/exposition.html.

52. Encarnación Pinedo, *Encarnación's Kitchen: Mexican Recipes from Nineteenth-Century California,* edited and translated by Dan Strehl (Berkeley: University of California Press, 1898), 19.

53. Ibid., 10.

54. Ibid., rolls, 61; cornbreads, 65–66; casseroles, 135–36; puddings, 186–87, 189–90; sopapillas, 183.

55. *NIRA Hearing on Code*, 1934, 5.

56. "Memorial of the American Baking Powder Association," BPC, 64.

Chapter 6. The Rise of Baking Powder Business: The Midwest

1. Reid Badger, "World's Fairs and Other Recipes for Progress," introduction to *Favorite Dishes: A Columbian Autograph Souvenir Cookery Book*, compiled by Carrie V. Shuman (Urbana: University of Illinois Press, 2001), xxvi.

2. Marilyn Kern-Foxworth, *Aunt Jemima, Uncle Ben, and Rastus: Blacks in Advertising, Yesterday, Today, and Tomorrow* (Westport, CT: Praeger, 1994), 64.

3. M. M. Manring, *Slave in a Box: The Strange Career of Aunt Jemima* (Charlottesville: University of Virginia Press, 1998), 64, 72.

4. Deborah Cadbury, *Chocolate Wars: The 150-Year Rivalry between the World's Greatest Chocolate Makers* (New York: Public Affairs, 2010), 79. In 1875 Nestlé combined his powdered milk with chocolate and created "the world's first ready-made milk chocolate drink" (82).

5. Manring, *Slave in a Box*, 73–75.

6. Memorandum of Agreement between RCW and Alfred Bannister, Nov. 20, 1891, 2, RCW. A non-competition clause is also in an agreement dated July 21, 1900, between RCW and Messrs. Woods, Mailliard, and Schmiedell of San Francisco, California.

7. Carl Stephens, ed. *The Alumni Record of the University of Illinois* (Dixon, IL: Rogers Printing Co., 1921), 357, 359.

8. "Calumet County History," Calumet County, WI, website, http://www.co.calumet.wi.us/index.aspx?NID=182.

9. "Our Township," Charter Township of Calumet, MI, website, http://calumettownship.org/about.php.

10. John Licinius Everett Peck, Otto Hillock Montzheimer, and William J. Miller, *Past and Present of O'Brien and Osceola Counties, Iowa*, vol. 1 (Indianapolis: B. F. Bowen & Co., 1914), 442.

11. "Calumet City History," Calumet City, IL, website, http://calumetcity.org/history-2.

12. Ann Hagedorn Auerbach, *Wild Ride: The Rise and Tragic Fall of Calumet Farm, Inc., America's Premier Racing Dynasty* (New York: Henry Holt, 1995), 36.

13. Richard Tedlow, *Giants of Enterprise: Seven Business Innovators and the Empires They Built* (New York: HarperCollins, 2001), 15.

14. Courtesy of Kraft Foods, Inc., email to the author, Feb. 9, 2010.

15. Advertisement, *Ann Arbor Register*, Nov. 21, 1895, 1.

16. "History," Wabash River, http://www.wabashriver.us/history/index.htm.

17. Historic marker, NW corner of Seventh Street and Wabash Avenue, Terre Haute, IN, http://www.waymarking.com/waymarks/WM1F5Q_Crossroads_of_America_1998_Marker.

18. Daniel Walker Howe, *What Hath God Wrought: The Transformation of America, 1815–1848* (Oxford: Oxford University Press, 2007), 141, 136.

19. Benj. F. Stuart, *History of the Wabash and Valley* (n.p.: Longwell-Cumings Co., 1924), Rothschilds, 48; Irish, 43.

20. Howe, *What Hath God Wrought*, 218.

21. Ibid., 255.

22. Ibid., 136–41.

23. Ibid., 140.

24. A. R. Markle and Gloria M. Collins, *The House of Hulman: A Century of Service, 1950–1950* (Terre Haute, IN: Clabber Girl, 1952), 31.

25. Ibid., 31.

26. Ibid.

27. Ibid., 15.

28. Ibid., 16–17.

29. Harvey L. Carter, "Rural Indiana in Transition, 1850–1860," *Agricultural History* 20, no. 2 (1946): 107–108.

30. Ibid., 112.

31. Ibid., 110.

32. Markle and Collins, *House of Hulman*, 34.

33. Ibid., 44–48.

34. "German Immigration," *U.S. Immigration and Migration Reference Library*, edited by Lawrence W. Baker, et al., vol. 1: Almanac, U*X*L, 2004, 221–46), U.S. History in Context, http://ic.galegroup.com.

35. "German American Culture," at ibid.

36. "The Germans in America Chronology," Library of Congress, http://loc.gov/rr/european/imde/germchro.html.

37. William Ferguson, *America by River and Rail* (London, 1856), 347, in Carter, "Rural Indiana in Transition," 108.

38. Markle and Collins, *House of Hulman*, 38.

39. Ibid., 48–49.

40. Ibid., 67.

41. Ibid., 73.
42. Ibid., 67.
43. Ibid.,106–107, 114.
44. Ibid., 97.
45. Ibid., 94–96.
46. Ibid., 95.
47. Ibid., 94–97.
48. Ibid., 90.
49. Ibid., 122.
50. Ibid., 107.
51. Ibid., 119.
52. Ibid., 178.
53. Ibid., 189.
54. "Eugene V. Debs: American Social and Labour Leader," *Encyclopaedia Brittanica*, http://www.britannica.com/EBchecked/topic/154766/Eugene-V-Debs.
55. "History of the Federal Judiciary: The Debs Case: Labor, Capital, and the Federal Courts of the 1890s, Biographies, American Railway Union," *Federal Judicial Center*, http://www.fjc.gov/history/home.nsf/page/tu_debs_bio_aru.html.
56. "Biographical Sketch," Eugene V. Debs Papers, Manuscript and Visual Collections Department, William Henry Smith Memorial Library, Indiana Historical Society, Collection #SC 0493, http://www.indianahistory.org/our-collections/collection-guides/eugene-v-debs-papers-1881-1940.pdf/?searchterm=debs.
57. "Debs Biography," http://debsfoundation.org/index.php/landing/debs-biography.
58. Markle and Collins, *House of Hulman*, 171–72.
59. Herman Hulman Letter to Eugene V. Debs, Oct. 13, 1893, Wabash Valley Visions and Voices Digital Memorial Project, Eugene V. Debs Collection, http://visions.indstate.edu:8888/cdm/ref/collection/evdc/id/4511.
60. Markle and Collins, *House of Hulman*, 194.
61. Ibid., 81–82.
62. Ibid., 198–99.
63. Henriette Davidis, *Pickled Herring and Pumpkin Pie: A Nineteenth-Century Cookbook for German Immigrants to America*, introduction by Louis A. Pitschmann, Monographs of the Max Kade Institute (Madison, WI: Max Kade Institute for German-American Studies, 2003), xi-xii.
64. Ibid., 463.
65. Ibid., cakes, 489–97; fritters, shortcake, griddle cakes, 486–87.
66. Ibid., cake instructions, 371; biscuits, 390–91; cup cake, 392; Portuguese, 394.
67. Ibid., "Speculaci," 402; "Hohenzollern Cakes," 404; *pfeffernuesse*, 406; honey cakes, 407–408.
68. Ibid., 371.
69. Paul R. Jones, "Justus von Liebig, Eben Horsford, and the Development of the Baking Powder Industry," *Ambix* 40, part 2 (July 1993): 73.
70. "Dr. Oetker Baking Powder," Dr. Oetker, Our Products, Baking Ingredients, http://www.oetker.ca/ca-en/our-products/dr-oetker-baking-powder/dr-oetker-baking-powder.html.
71. RCW, series A, vol. B, Stockholders' and Directors' Record Book B, 1883–1906, Oct. 20, 1893, 121.
72. "Prof. Horsford Dead," *Cambridge Tribune* 15, no. 44, Jan. 7, 1893, 3.

73. "In Memoriam, Eben Norton Horsford, July Twenty-Seventh, 1813. January First, 1893," 30, Wellesley College Library, https://archive.org/details/inmemoriamebennooowell.

74. Woman's Christian Temperance Union letter "To Our Friends," Feb. 12, 1898, RCW.

75. Auerbach, *Wild Ride*, 37.

76. Ibid., 37–38.

77. Markle and Collins, *House of Hulman*, 199.

78. Ibid., 200.

Chapter 7. The Pure Food War

1. "Reconstruction Politics in Missouri," *American Experience*, PBS, http://www.pbs.org/wgbh/americanexperience/features/general-article/james-politics.

2. "Thomas Theodore Crittenden," National Governors Association, http://www.nga.org/cms/home/governors/past-governors-bios/page_missouri/col2-content/main-content-list/title_crittenden_thomas.html.

3. Ted P. Yeatman, *Frank and Jesse James* (Nashville, TN: Cumberland House Publishing, 2000), meeting, 266; pardon, 275; amnesty bill, 363.

4. *Laws of Missouri, Fortieth General Assembly* (Jefferson City, MO: Tribune Printing Co., State Printers and Binders, 1899), S.B. 887.

5. "A Lead Pipe Cinch," *Kansas City Packer*, Sept. 2, 1899, in BPC, 1966.

6. Claude Wetmore, *The Battle against Bribery* (St. Louis: Pan-American Press, 1904), 146.

7. Report of the Secretary, ABPA, 1900, BPC, 967.

8. Membership form, BPC, 1591.

9. Introduction, BPC, 6.

10. Hearing U. S. Senate Committee on Agriculture and Forestry, Jan. 8, 1900, BPC, 136.

11. Report of the Secretary, ABPA, Oct. 16, 1900, BPC, 977.

12. Morrison, Report of the Secretary, Oct. 29, 1901, BPC, 985.

13. Richard Tedlow, *Keeping the Corporate Image: Public Relations and Business, 1900–1950, Industrial Development and the Social Fabric*, vol. 3 (Greenwich, CT: JAI Press, 1979), 2, 4–5, 26.

14. Introduction, BPC, 7; emphasis in the original.

15. Morrison, Report of the Secretary, Oct. 16, 1900, BPC, 976–77.

16. "Colonel Huffman Indicted," *Indianapolis Journal*, April 12, 1901, 8.

17. Morrison, Report of the Secretary, Oct. 29, 1901, BPC, 984–85.

18. "The Public Deceived by Literature Entitled 'Alum Baking Powder,' Which Is Inspired and Paid for, etc.," n.d., BPC, 1635.

19. "Advertisements Issued by Baking Powder Companies," Jan. 1901, BPC, 1636–38.

20. "Royal Baking Powder Trust Turns Tail and Flees," and "Trust Secretly Starts Another Baking Powder Fight in Minnesota," in *Grocery World*, n.d., BPC, 863–66.

21. "Trust Secretly Starts Another Baking Powder Fight in Minnesota," *Grocery World*, BPC, 866.

22. "A New Drug Habit," Hearing on House Bill 1032, Commonwealth of Massachusetts, Feb. 5, 1903, BPC, 1068.

23. *State of Missouri v. Whitney Layton*, St. Louis Court of Criminal Correction, City of St. Louis, Missouri, Defendant's Brief, BPC, 675, citing *People vs. Marx*, 99 N.Y. 377.

24. Ernest Ellsworth Smith, *Aluminum Compounds in Food* (New York: Paul B. Hoeber, 1928), 154.

25. Ibid., 159.

26. Letter from attorneys Seddon & Blair to Patterson Bain, Dec. 8, 1900, BPC, 1355–56.

27. *State of Missouri v. Whitney Layton*, St. Louis Court of Criminal Correction, City of St. Louis, BPC, 703–704.

28. *State of Missouri v. Whitney Layton*, Missouri Supreme Court, Oct. 1900, BPC; *Mugler*, 742; affirmed, 782.

29. *Whitney Layton, Plff. in Err., v. State of Missouri*; *Whitney Layton v. State of Missouri*, 187 U.S. 356 (1902), Docket No. 69; Parallel Citations, 187 U.S. 356; 23 S.Ct. 137; 47 L.Ed. 214.

30. U.S. Congress, Senate, Committee on Agriculture and Forestry, Pure Food Legislation: "Memorial of the American Baking Powder Association in the Matter of the Bill S. 3618," April 21, 1900, presented by Mr. Proctor, Committee on Agriculture and Forestry, BPC, 61.

31. *Bureau of Labor Statistics, Eleventh Annual Report*, Missouri State Archives, Nov. 5, 1889 (Jefferson City, MO: Tribune Printing Co., 1889), 14.

32. Ibid., 289–90, 297. KC, an alum baking powder, was ten cents per pound at this time.

33. "About FDA, What We Do, History—Part I, The 1906 Food and Drugs Act and Its Enforcement," U.S. Food and Drug Administration, http://www.fda.gov/AboutFDA/WhatWeDo/History/Milestones/ucm128305.htm.

34. U.S. Congress, "Memorial of the American Baking Powder Association," BPC, 52.

35. Ibid., BPC, 65.

36. Testimony by Mr. Higgins, president of the ABPA, "Memorial of the American Baking Powder Association," BPC, 120.

37. Ibid., BPC, 133–34.

38. Affidavit of D. J. Kelley, BPC, 694, in U.S. Congress, *Congressional Serial Set* (Washington, DC: U.S. Government Printing Office, 1901), 694–700.

39. Affidavit of Evelyn B. Baldwin; Nov. 21, 1903, BPC, 1265, in ibid., 691–92.

40. Morrison, Association Report of the Secretary, 1900–1901, BPC, 980.

41. "The Great American Lobby: The Typical Example of Missouri," *Leslie's Popular Monthly* 56 (August 1903): 382–93, BPC, 32.

42. Affidavit of John A. Lee, BPC, 1263, in U.S. Congress, *Congressional Serial Set*.

43. Wetmore, *Battle against Bribery*, BPC, 1811.

44. Ibid., BPC, 1813.

45. "Senate Tie on the Alum Bill Decided by Lieut. Gov. Lee," *St. Louis Globe-Democrat*, March 5, 1903, BPC, 1856.

46. "History of the Alum Bill," *The Times*, Kansas City, MO, April 8, 1903, BPC, 1866–67.

47. Chronological Synopsis, BPC, 1869.

48. "Age and Sex Composition," in Frank Hobbs and Nicole Stoops, U.S. Census Bureau, Census 2000 Special Reports, Series CENSR-4, *Demographic Trends in the 20th Century* (Washington, DC: U.S. Government Printing Office, 2002), https://www.census.gov/prod/2002pubs/censr-4.pdf.

49. Harold S. Wilson, *McClure's Magazine and the Muckrakers* (Princeton, NJ: Princeton University Press, 1970), 28.

50. Peter Lyon, *Success Story: The Life and Times of S. S. McClure* (New York: Charles Scribner's Sons, 1963), 114.

51. Robert Cantwell: "Journalism—the Magazines," in *The Muckrakers and American Society*, edited by Herbert Shapiro (Lexington, MA: D. C. Heath and Co., 1968), 21–22.

52. Lincoln Steffens, *The Letters of Lincoln Steffens*, vol. 1: *1889–1919* (New York: Harcourt, Brace and Co., 1938), 164.

53. Peter Hartshorn, *I Have Seen the Future: A Life of Lincoln Steffens* (Berkeley: Counterpoint, 2011), 114.

54. Upton Sinclair, "The Muckrake Man," in Shapiro, *Muckrakers and American Society*, 12.

55. Wilson, *McClure's*, 24.

56. Ibid., 26.

57. Ibid., 23–24.

58. Wetmore, *Battle against Bribery*, foreword, BPC, 1803.

59. Hartshorn, *I Have Seen the Future*, 115.

60. Affidavit of D. J. Kelley, BPC, 1568.

61. "I Boodle? No! Says Stone," *Chicago Daily Tribune*, April 21, 1903.

62. The Missouri legislature elected Stone to the U.S. Senate in 1902 and 1908. The people of Missouri elected him in 1914. "Stone, William Joel," *Biographical Directory of the United States Congress*, http://bioguide.congress.gov/scripts/biodisplay.pl?index=S000968. During his tenure in the U.S. Senate, Stone missed 34.4 percent of roll call votes, well above the average of 27.2 percent missed by senators in 1918. "Sen. William Stone," Govtrack, https://www.govtrack.us/congress/members/william_stone/410427.

63. Lee letter to Kelley, April 7, 1903, BPC, 1830.

64. Wetmore, *Battle against Bribery*, 179.

65. Ibid., BPC, 1817.

66. Kelley-Lee Correspondence from *St. Louis Post-Dispatch*, Jan. 3, 1903, BPC, 1834.

67. Ibid., n.d., BPC, 1835.

68. Wetmore, *Battle against Bribery*, BPC, 1819.

69. Chronological Synopsis of Bribery Investigations and Trials Resulting from Missouri Anti-Alum Legislation, 1903–1905, BPC, 1870.

70. "Lee Has Vanished: Lieutenant Governor of Missouri Wanted," *Los Angeles Herald*, April 18, 1903, 2.

71. "Lieut. Governor Lee Central Figure in Missouri Hunt," *New York Herald*, April 19, 1903, BPC, 1889.

72. Chronological Synopsis, BPC, 1870–71.

73. Ziegler Indictment, BPC, 1216.

74. Chronological Synopsis, BPC, 1871.

75. "Seven Indictments Returned by Cole County Grand Jury," *St. Louis Republic*, April 19, 1903, BPC, 1887.

76. "Daniel J. Kelley, Reported to Have Fled to London," *St. Louis Republic*, March 1, 1904, BPC, 1942.

77. In the Matter of Requisition for William Ziegler, as a fugitive from justice, BPC, earnest, 1280; commonwealth, 1280.

78. "Ziegler Weeps under Fire of Accusation," *New York American*, Dec. 3, 1903, BPC, 1938.

79. In the Matter of Requisition for William Ziegler, BPC, 1343.

80. "Court Revokes Governors' Right to Deny Extradition," *Los Angeles Times*, June 24, 1987.

81. Cornelius C. Regier, *The Era of the Muckrakers* (Chapel Hill: University of North Carolina Pres, 1932), 6.

82. Lincoln Steffens, "Enemies of the Republic," BPC, 16.

83. Lincoln Steffens, *Autobiography* (New York: Harcourt, Brace and Co., 1931), 447–48.

84. *Ziegler v. Hoagland, et al.*, May 28, 1888, BPC, 614.

85. It was located at 26 Ridge Road in Summit, New Jersey. Antoinette Martin, "Bygone Era, for Sale in Summit," *New York Times*, July 21, 2011, http://www.nytimes.com/2011/07/24/realestate/a-bygone-era-for-sale-in-summit-in-the-regionnew-jersey.html?_r=0.

86. Report of the Secretary, Annual Meeting of the ABPA, Sept. 26, 1904, BPC, 1014-i.

87. Ibid., BPC, 1014-j.

88. E. Smith, *Aluminum Compounds in Food*, 162; *The People of the State of California v. J.A. Saint*, and *The People of the State of California v. A. Zimbleman*; both in Police Court of Los Angeles City, State of California, December 1903, BPC, 952–64.

89. In a case in Pennsylvania in 1910, the court ruled that baking powder *was* food.

90. *People of the State of California, v. J. A. Saint*; *People of the State of California, v. A. Zimbelman*.

91. A. Cressy Morrison, Address—"The Baking Powder Controversy," *Journal of Proceedings*, Eighth Annual Convention and International Pure Food Congress of the National Association of State Dairy and Food Departments, St. Louis, MO, Sept. 29, 1904, BPC, 1186. Another parallel might be that there were intimations that Standard Oil had bribed the Ohio state legislature. If it did, Rockefeller was not caught doing it personally.

92. Chronological Synopsis, BPC, 1868.

93. Decree, Office of the Governor, BPC, 2036. Folk took office on January 9, 1905.

94. Morrison, Introduction, BPC, 6.

95. "Jury Indicts Ziegler, Stone, and Luckett," *St. Louis Globe-Democrat*, Nov. 15, 1903, BPC, 1931–1933; Nov. 17, 1903, "Luckett Pleads Not Guilty," BPC, 1935. On November 14 the grand jury indicted Farris's attorney, F. E. Luckett, for jury tampering; however, Luckett was charged with only a misdemeanor. He had approached jurors and talked to them about how important it would be to have Farris acquitted, but he had not offered them a bribe.

96. Chronological Synopsis, BPC, 1871–73.

97. BPC, 30.

98. *Missouri Historical Review*, XI, Oct. 1916–July 1917, 108–11.

99. Pure Food and Drug Act, Section 7, 1906.

100. Report of the Secretary, APBA, Dec. 11, 1905, BPC, 1018.

101. "William Ziegler Dead after a Long Illness," *Evening Telegram*, New York, May 24, 1905, BPC, 1951–52.

102. Maggie Gordon, *The Gilded Age on Connecticut's Gold Coast: Transforming Greenwich, Stamford, and Darien* (Charleston, SC: History Press, 2014), ch. 12, "The Legacy of William Ziegler."

103. "Miss Brandt Loses Ziegler Will Case," *New York Times*, Oct. 6, 1913.

104. "Compromise Ends Suit to Test Ziegler Will," *New York Times*, Aug. 2, 1905, 7.

Chapter 8. The Alum War and World War I

1. Docket 540, FTC/Royal Respondent / Respondent's Brief (Royal's Brief) n.d., 10, http://babel.hathitrust.org/cgi/pt?id=mdp.39015073780168;view=1up;seq=11.

2. Eleventh Annual Report, BPC 32.

3. Minnesota Dairy and Food Report, BPC, 27.

4. BPC, 43, 41.

5. "Company History," American Maize-Products Co. History, http://www.fundinguniverse.com/company-histories/american-maize-products-co-history.

6. Ann Hagedorn Auerbach, *Wild Ride: The Rise and Tragic Fall of Calumet Farm, Inc., America's Premier Racing Dynasty* (New York: Henry Holt, 1995), 39.

7. "Biographical Note," Remsen (Ira) Papers, 1846–1927, ms. 39, Special Collections, Milton S. Eisenhower Library, Johns Hopkins University, http://ead.library.jhu.edu/ms039.xml.

8. D. J. Warner, "Ira Remsen, Saccharin, and the Linear Model," *Ambix* 55, no. 1 (2008): 50–61, U.S. National Library of Medicine, National Institutes of Health, http://www.ncbi.nlm.nih.gov/pubmed/18831154.

9. James Harvey Young, *Pure Food: Securing the Federal Food and Drugs Act of 1906* (Princeton, NJ: Princeton University Press, 1989), 68–71. The war between sugar and corn syrup continues in the twenty-first century.

10. "Report of Referee Board of Consulting Scientific Experts on Influence of Aluminum Compounds on the Nutrition and Health of Man" (Washington, DC: U.S. Department of Agriculture, 1914), in *Aluminum Compounds in Food* by Ernest Ellsworth Smith (New York: Paul B. Hoeber, 1928), 91–145.

11. Clayton A. Coppin and Jack High, *The Politics of Purity: Harvey Washington Wiley and the Origins of Federal Food Policy* (Ann Arbor: University of Michigan Press, 2002), 162.

12. Mark Pendergrast, *For God, Country, and Coca-Cola: The Definitive History of the Great American Soft Drink and the Company That Makes It* (New York: Basic Books, 2000), 116.

13. Harvey Washington Wiley, *1001 Tests of Foods, Beverages, and Toilet Accessories, Good and Otherwise* (New York: Hearst's International Library, 1914), xx.

14. Ibid.

15. Lorine Swainston Goodwin, *The Pure Food, Drink, and Drug Crusaders, 1879–1914* (Jefferson, NC: McFarland, 1999), 215. By 1925, *Good Housekeeping* had 1.5 million subscribers. Oscar E. Anderson Jr., *The Health of a Nation: Harvey W. Wiley and the Fight for Pure Food* (Chicago: University of Chicago Press, 1958), 264.

16. "The History of the Good Housekeeping Seal," Oct. 1, 2011, *Good Housekeeping*, http://www.goodhousekeeping.com/institute/about-the-institute/a16509/good-housekeeping-seal-history.

17. Ibid.

18. Wiley, *1001 Tests*, xx.

19. Ibid., xxvii.

20. Ibid.

21. Ibid., 3–4.

22. Calumet Baking Powder Company, *Why White of Eggs: A Truthful Sketch of the Inception and Efficiency of the Use of White of Eggs in Baking Powder; of the Recent Attempts to Restrict Its Use by False Interpretation of the Food Laws; and a Word as to the Future* (Chicago: Calumet Baking Powder Company, 1914), 33.

23. Ibid., 15.

24. Ibid., 15–16.

25. Ibid., 64–65.

26. Ibid., 62.

27. Ibid., 39.

28. Ibid., flyer, 41, 43.

29. Ibid., Kansas, 20; Idaho, 22, 24.

30. Ibid., 36.

31. Ibid.

32. "The Albumen Fraud," *American Food Journal* 9, no. 8 (1914): 341.

33. A. R. Markle and Gloria M. Collins, *The House of Hulman: A Century of Service, 1850–1950* (Terre Haute, IN: Clabber Girl, 1952), 220.

34. Ibid., 223–24.

35. "Musty Corn and the Dread Scourge Pellagra," July 14, 2015, Appalachian History Stories, Quotes, and Anecdotes, http://www.appalachianhistory.net/2015/07/musty-corn-and-dread-scourge-pellagra.html.

36. "Cornbread or Beaten Biscuits? Breaking the Food Code," Nov. 20, 2012, Appalachian History Stories, Quotes, and Anecdotes, http://www.appalachianhistory.net/2012/11/cornbread-or-beaten-biscuits-breaking-the-food-code.html.

37. Minnie C. Fox, comp., *The Blue Grass Cook Book,* new introduction by Toni Tipton-Martin (Lexington: University of Kentucky Press, 2005), 1–2.

38. Tamera Alexander, "Beaten Biscuits, a Southern Tradition," Oct. 24, 2012, http://tameraalexander.blogspot.com/2012/10/beaten-biscuits-southern-tradition.html.

39. Mrs. Simon Kander, comp., *The "Settlement" Cook Book: Containing Many Recipes Used in Settlement Cooking Classes, the Milwaukee Public School Cooking Centers, and Gathered from Various Other Reliable Sources* (Milwaukee, 1901), 28.

40. Ibid., 44.

41. Ibid., biscuits, 39; strawberry shortcake, 43; griddle cakes, 44–46; French, 55; German, 56; potato, 57.

42. Ibid., 213.

43. Ibid., 82–83.

44. Ibid., 179.

45. Ibid., 301–302, 322.

46. Ibid., "Suet," 258; "Steamed Fruit," 260; "Chocolate Pudding No. 1," 261; "Steamed Chocolate," 263.

47. Ibid., 347.

48. Ibid., 350, 346.

49. Ibid., 354; "Chocolate Zweiback," 357; "Matzohs," 362; "Poppyseed," 364; "Rye," 365.

50. Ibid., anise, 377; pfeffernuesse, 378, gingerbread, 381, lebkuchen, 382; peanut, 372, chocolate, 374, oatmeal, 375.

51. William Woys Weaver, *The Christmas Cook: Three Centuries of American Yuletide Sweets* (New York: HarperPerennial, 1990), 101–102, icing, 129.

52. Ibid., 119.

53. Ibid., 9.

54. Ibid., 127.

55. Barbara Swell, *The 1st American Cookie Lady* (Asheville, NC: Native Ground Music, 2005).

56. Ibid., 86–88.

57. Ibid., 2, 129.

58. Ibid., recipes 106, 153; pp. 39, 41.

59. Celia M. Kingsbury, "'In Close Touch with Her Government': Women and the Domestic Science Movement in World War One Propaganda," in *The Recipe Reader: Narratives, Contexts, Traditions*, edited by Janet Floyd and Laurel Forster (2003; Lincoln: University of Nebraska Press, 2010), 92, 93, 97–98.

60. Swell, *1st American Cookie Lady*, 142.

61. "Eggless, Milkless, Butterless Cake," Royal Baking Powder Company, *55 Ways to Save Eggs*, 1917, 10, Duke University Libraries Digital Collections, http://library.duke.edu/digital collections/eaa_CK0072.

62. Paul R. Mullins, *Glazed America: A History of the Doughnut* (Gainesville: University Press of Florida, 2008), 29–30.

63. Michael Krondl, "The War That Made Donuts a Secret Weapon," June 4, 2014, http://zesterdaily.com/cooking/salvation-army-doughnuts-in-great-war.

64. "The Royal Baking Powder Company versus the Montana State Board of Health," *American Food Journal* (June 1920): 13–14.

Chapter 9. The Federal Trade Commission Wars

1. *The Metropolitan Life Cook Book* ([n.p.] Metropolitan Life Insurance Company, 1922), 3.

2. "Royal Baking Powder Had Record Sale in 1921," *American Food Journal* 17 (Jan. 1922).

3. *NIRA Hearing on Code of Fair Practices and Competition Presented by the Baking Powder Industry*, April 30, 1934, 83.

4. Carolyn M. Goldstein, *Creating Consumers: Home Economists in Twentieth-Century America* (Chapel Hill: University of North Carolina Press, 2012), 186.

5. Ernest Ellsworth Smith, *Aluminum Compounds in Food* (New York: Paul B. Hoeber, 1928), 167–68.

6. *Alum in Baking Powder: The Complete Text of the "Trial Examiner's Report Upon the Facts" . . . in the Matter of Royal Baking Powder Company, Docket No. 540* (New York: Royal Baking Powder Company, 1927), 1.

7. "Baking Powder Hearing," *Wall Street Journal*, Nov. 24, 1927, 13; Anne Hagedorn Auerbach, *Wild Ride: The Rise and Tragic Fall of Calumet Farm, Inc., America's Premier Racing Dynasty* (New York: Henry Holt, 1995), 41.

8. *The Truth about Baking Powder*, compiled and distributed by Calumet Baking Powder Company (Chicago, 1928), 18.

9. "News Notes," *National Grocers Bulletin*, Sept. 1929, 69.

10. Docket 540, FTC/Royal Respondent / Respondent's Brief (Royal's Brief) n.d., 20.

11. Ibid., 3.

12. *Alum in Baking Powder*, 80.

13. A. R. Markle and Gloria M. Collins, *The House of Hulman: A Century of Service, 1850–1950* (Terre Haute, IN: Clabber Girl, 1952), 240.

14. Ibid., 241.

15. Ibid., 242.

16. Ibid., 254.

17. Ibid.

18. Martha H. Patterson, *Beyond the Gibson Girl: Reimagining the American New Woman, 1895–1915* (Urbana: University of Illinois Press, 2005), 2.

19. James D. Norris, *Advertising and the Transformation of American Society, 1865–1920* (New York: Greenwood Press, 1990), 106.

20. Frank Presbrey, *The History and Development of Advertising* (Garden City, NY: Doubleday, Doran & Co., 1929), 442.

21. Ibid., 443.

22. Michigan State University Libraries, Gerald M. Kline Digital and Multimedia Center. https://archive.lib.msu.edu/DMC/sliker/msuspcsbs_bake_walterbake49/msuspcsbs_bake_walterbake49.pdf.

23. Markle and Collins, *House of Hulman*, 255.

24. Ibid., 259.

25. Kewpie: "Bonniebrook Homestead, Taney County, Missouri," National Park Service, Women's History Month, http://www.nps.gov/nr/feature/wom/2008/bonniebrook-homestead.htm; Grace Drayton: "Drayton, Grace Gebbie (1877–1936)," *Women in World History: A Biographical Encyclopedia*. *Encyclopedia.com.* http://www.encyclopedia.com/women/encyclopedias-almanacs-transcripts-and-maps/drayton-grace-gebbie-1877-1936; on the Katzenjammer Kids, see "Rudolph Dirks, American Cartoonist," *Encyclopaedia Brittanica*, http://www.britannica.com/EBchecked/topic/165121/Rudolph-Dirks.

26. "The Trust," *Reliable Recipes* (Chicago: Calumet Baking Powder Company, 1909), cover; "A Trust," *Calumet Cook Book* (Chicago: Calumet Baking Powder Company, ca. 1923), cover.

27. "Buster Brown Bread (1907)," University Libraries, University of Washington Digital Collections, digitalcollections.lib.washington.edu/cdm/ref/collection/advert/id/448. Buster and Tige as advertising icons for Buster Brown Shoes began in 1904. "Brown Shoe Company History," Funding Universe, http://www.fundinguniverse.com/company-histories/brown-shoe-company-inc-history. They persisted into the 1950s, accompanied by a jingle: "Does your shoe have a boy inside? What a funny place for a boy to hide."

28. Frances Hodgson Burnett, *Little Lord Fauntleroy*, 2nd ed. (New York, 1889).

29. Robert Brauneis, "Copyright and the World's Most Popular Song," Oct. 14, 2010. 56 *Journal of the Copyright Society of the U.S.A.* 335 (2009), GWU Legal Studies Research Paper No. 392.

30. Sinclair Lewis, *Elmer Gantry*, afterword by Mark Schorer (1927; New York: Signet Classic, 1970), 351.

31. Ibid., 420.

32. KKK Flyer, Vigo County Historical Society.

33. *The Birth of a Nation*, directed by D. W. Griffith, David W. Griffith Corp, 1915.

34. Kathleen M. Blee, *Women of the Klan: Racism and Gender in the 1920s* (Berkeley: University of California Press, 1991), 94.

35. "Ku Klux Klan in Indiana," Indiana State Library, http://www.in.gov/library/2848.htm.

36. Blee, *Women of the Klan*, 108.

37. Leonard J. Moore, *Citizen Klansmen: The Ku Klux Klan in Indiana, 1921–1928* (Chapel Hill: University of North Carolina Press, 1997), 58.

38. Ibid., 60.

39. "Terre Haute Pays Tribute to Victim of Lawless Conditions," *Fiery Cross*, Feb. 1, 1924, 1. Detective Steve Kendall was killed during a gas station robbery. There was no indication that Kendall was a member of the Klan, but his death was attributed to "aliens." "Terre Haute Pays Tribute to Second Murdered Klansman," *Fiery Cross*, Feb. 8, 1924, 1. John Smith was a grocery store owner and Klan member. Moore, *Citizen Klansmen*, 25.

40. "Terre Haute Pays Tribute to Victim," 5.

41. Blee, *Women of the Klan*, 127–28.

42. Ibid., 147–51.

43. Ibid., 157–58.

44. "Red Light District," Historic Events, Vigo County Public Library, http://www.vigo.lib.in.us/subjects/genealogy/timeline/historical.

45. "Terre Haute Is Shocked," *Fiery Cross*, March 28, 1924, 1.

46. Ibid.

47. David J. Goldberg, *Discontented America: The United States in the 1920s* (Baltimore: Johns Hopkins University Press, 1999), 135.

48. "Great Atlantic & Pacific Tea Company, Inc. (A&P)," *Encyclopaedia Brittanica*, https://www.britannica.com/topic/Great-Atlantic-and-Pacific-Tea-Company-Inc.

49. Marc Levinson, *The Great A&P and the Struggle for Small Business in America* (New York: Hill and Wang, 2011), 38–40. However, in *New and Improved*, Richard Tedlow quotes John A. Hartford, "the younger son of A&P owner George Huntington Hartford," as saying, "I think it was about '90 or '91, [that we] first got into baking powder" (190). *New and Improved: The Story of Mass Marketing in America* (Boston: Harvard Business School Press, 1996).

50. Lewis Corey, *The House of Morgan: A Social Biography of the Masters of Money* (New York: G. Howard Watt, 1930), 442.

51. Tedlow, *New and Improved*, 218–19.

52. Ibid., $1 billion, 10; chain stores, 182; 15,418 stores, http://aptea.com/our-company/our-history.

53. Norris, *Advertising and Society*, 97.

54. Markle and Collins, *House of Hulman*, 244–45.

55. Ibid., 247.

56. Ibid.

57. "Baking Powder and Kidneys," Ramsay Spillman, MD, letter to the editor, *Journal of the American Medical Association* 89, no. 17 (1927):1445, http://jama.ama-assn.org/cgi/content/summary/89/17/1445-b.

58. Presbrey, *History and Development of Advertising*, 584, http://institutionalmemory.hbs.edu/timeline/1927/william_ziegler_jr_gives_1_million_for_international_research_chair.html.

59. "David Will Leave Business School, To Serve as Executive Vice President of Royal Baking Co—Was in Navy during War," *Harvard Crimson*, June 2, 1927, http://www.thecrimson.com/article/1927/6/2/david-will-leave-business-school-pthe/?print=1.

60. "Development and Use of Baking Powder and Baking Chemicals," United States Department of Agriculture, Circular No. 138, Washington, DC, Nov. 1930, 5.

61. Auerbach, *Wild Ride*, 43.

62. Corey, *House of Morgan*, 442.

63. Ibid.

64. "'Trick' Selling Condemned by Federal Trade Commission," Sales Management, July 6, 1929, 20; Chamber of Commerce statement to members, June 21, 1929.

65. Then, chicken was an expensive, prestigious food.

Chapter 10. The Price War

1. The original Toll House recipe called for 1 teaspoon of baking soda dissolved in 1 teaspoon of hot water. Many recipes now use baking powder. "Toll House Cookies Are Easy to Make," *Brooklyn Daily Eagle*, May 24, 1940, 6, http://bklyn.newspapers.com/clip/3999204/1940_toll_house_chocolate_chip_cookie.

2. Irma S. Rombauer, *The Joy of Cooking*, facsimile of first ed., foreword by Edgar Rombauer (1931; New York: Scribner, 1998), n.p. (fourth page of foreword).

3. Ruth Adams Bronz, "What's Cooking?" *New York Times*, March 23, 1997, http://www.nytimes.com/books/97/03/23/reviews/970323.23bronzt.html.

4. Rombauer, *Joy of Cooking*, 187–282.

5. Mendelson, *Stand Facing the Stove*, 161–2.

6. Irma S. Rombauer, illustrations by Marion Rombauer Becker, *The Joy of Cooking* (Indianapolis: Bobbs-Merrill, 1943), 447–48.

7. Ibid., 450.

8. James Gray, *Business without Boundary: The Story of General Mills* (Minneapolis: University of Minnesota Press, 1954), 211–13.

9. *Betty Crocker's 101 Delicious Bisquick Creations as Made and Served by Well-Known Gracious Hostesses, Famous Chefs, Distinguished Epicures, and Smart Luminaries of Movieland*, ([n.p.] General Mills, 1933), back cover.

10. Gray, *Business without Boundary*, 213.

11. *101 Delicious Bisquick Creations*, 6–7.

12. Ibid., 5.

13. Ibid., inside back cover.

14. Ibid., 5.

15. *How to Take a Trick a Day with Bisquick*, as told to Betty Crocker ([n.p.] General Mills, 1935), 31.

16. Gray, *Business without Boundary*, 210.

17. *How to Take a Trick a Day*, 23.

18. Ibid., 13.

19. Paul Lukas, "Jiffy's Secret Recipe Chelsea Milling has beaten its competition—the Pillsbury Dough Boy and Betty Crocker never laid a whisk on 'em—and plans to stay on top with two parts aw-shucks family business and one part professional management," CNN Money, Dec. 1, 2001, http://money.cnn.com/magazines/fsb/fsb_archive/2001/12/01/315113.

20. Advertisement in *Ladies' Home Journal*, April 1930.

21. A. R. Markle and Gloria M. Collins, *The House of Hulman: A Century of Service, 1850–1950* (Terre Haute, IN: Clabber Girl, 1952), 270.

22. Sigur E. Whitaker, *Tony Hulman: The Man Who Saved the Indianapolis Motor Speedway* (Jefferson, NC: McFarland, 2014), ch. 6, "Tony's Early Adventures."

23. Markle and Collins, *House of Hulman*, 252.

24. Ibid., 253.

25. Richard Tedlow, *New and Improved: The Story of Mass Marketing in America* (Boston: Harvard Business School Press, 1996), 183.

26. Ibid., 18.

27. "Let's Go Way Back," Hostess, http://www.hostesscakes.com/about.

28. Markle and Collins, *House of Hulman*, 272.

29. "This Is Your Guaranty," *Good Housekeeping* 90, no. 1 (1930): 8.

30. *Clabber Girl: The Healthy Baking Powder* (Terre Haute, IN: Hulman & Co., 1931), cover.

31. *Clabber Girl Baking Book* (Terre Haute, IN: Hulman & Co., ca. 1932).

32. RCW Box 1b, Statements 1932 & Annual Meeting, Feb. 8, 1933.

33. RCW, Report of Treasurer, Feb. 13, 1935, 2.

34. RCW, Report of Treasurer, Feb. 12, 1936, "A."

35. RCW, Report of Treasurer, Feb. 12, 1936, 3.

36. RCW Box 1b, Statements 1932 & Annual Meeting, Feb. 8, 1933.

37. *24 Uses for Rumford Phosphate Baking Powder in Cooking 64 New Uses for Rumford All-Phosphate Baking Powder in Daily Cooking* (Providence, RI: The Rumford Company, 1929); *64 New Uses for Rumford All-Phosphate Baking Powder in Daily Cooking* (Providence, RI: The Rumford Company, 1935).

38. RCW Box 1b, Statements 1932 & Annual Meeting, Feb. 8, 1933, 5–6.

39. RCW, Statements 1933, 4.

40. RCW Box 1b, Statements 1932 & Annual Meeting, Feb. 8, 1933, 2.

41. RCW, Report of Treasurer, Feb. 13, 1935, 3.

42. Vintage Ad, "KC Baking Powder Grocer's Want Book," Jaques Mfg. Co., Chicago IL, Seller information kds99 (6921).

43. RCW Box 1b, Statements 1934 & Annual Meeting, Feb. 13, 1935.

44. "Baking Powder," *Bakers Weekly*, May 27, 1933, 57–59.

45. "Won't Standardize Baking Powder Cans," *American Food Journal* 17 (Oct. 1922): 46.

46. NIRA Code, 1934, 5.

47. RCW Statements 1933 & Annual Meeting, 1934, 2.

48. NIRA Code, 1934, 5.

49. Ibid., 88.

50. Ibid., 87.

51. Ibid.

52. Ibid., 88.

53. Ibid., 94.

54. Ibid., 110.

55. Ibid., 204.

56. Ibid., 214.

57. "The Department in the New Deal and World War II 1933–1945," United States Department of Labor, http://www.dol.gov/dol/aboutdol/history/dolchp03.htm.

58. RCW, series A, box 2, file 55.

59. RCW Report to Shareholders, 1934, 3.

60. *The Little Book of Excellent Recipes*, by The Mystery Chef, Davis Baking Powder cookbook, 1934.

61. Courtesy of Kraft Foods, Inc., email to the author, Feb. 9 2010.

62. RCW, Report of Treasurer, Feb. 10, 1937, 1.

63. Ibid., 2.

64. Ibid., 3.

65. Markle and Collins, *House of Hulman*, 292–93.

66. Ibid., 286–87.

67. Eva Brunson Purefoy, *The Purefoy Hotel Cook Book* (Talladega, AL, 1938); Marie Goebel Kimball, *The Martha Washington Cook Book* (New York: Coward-McCann, 1940); Mrs. S. R. Dull, illustrated by Lucina Wakefield, *Southern Cooking* (New York: Grosset & Dunlap, 1941); Helen Duprey Bullock and William Parks, *The Williamsburg Art of Cookery; or, Accomplish'd Gentlewoman's Companion: Being a Collection of Upwards of Five Hundred of the Most Ancient & Approv'd Recipes in Virginia Cookery . . . And Also a Table of Favorite Williamsburg Garden Herbs* (Colonial Williamsburg, VA, 1942).

68. *Gone with the Wind Cook Book: Famous "Southern Cooking" Recipes* (New York: Anthony P. Cima, 1940), 21.

69. Lillie S. Lustig, S. Claire Sondheim, and Sarah Rensel, comp. and ed. *The Southern Cook Book of Fine Old Recipes* (Reading, PA: Culinary Arts Press, 1939), Mammy, 33; Dinah, 34; Confederate, Pickaninny, 42.

70. The movie studios tread lightly around the New Deal, which was investigating them for monopolistic practices.

71. *Gone with the Wind Cook Book*, 21–26.

72. Vivian May, "Cross Purposes," *Southern Changes* 16, no. 2 (1994): 13–17.

73. "Idella's Crisp Biscuits from Cross Creek Cookery by Marjorie Kinnan Rawlings," March 30, 2015, Hub Pages, http://pstraubie48.hubpages.com/hub/Idellas-Crisp-Biscuits-from-Cross-Creek-Cookery-by-Marjorie-Kinnan-Rawlings.

74. William G. Panschar, *Baking in America: Economic Development*, vol. 1 (Evanston, IL: Northwestern University Press, 1956), 227–28.

75. Ibid., 227.

76. Anne Mendelson, *Stand Facing the Stove: The Story of the Women Who Gave America* The Joy of Cooking (New York: Henry Holt, 1996), 172.

77. Panschar, *Baking in America*, 237.

78. *Royal Cakes* (New York: Standard Brands Inc., 1950), 2–3; italics in original.

79. Ibid., 10.

80. Ibid.

81. Markle and Collins, *House of Hulman*, 317.

82. Ibid., 324.

Chapter 11. Baking Powder Today

1. A search of WorldCat produced no cookbooks from Royal after 1957. *Here Are the Cakes America Loves: Royal Cakes Made with Royal Cream of Tartar Baking Powder* (New York: Standard Brands Inc., 1950).

2. James Beard, *The New James Beard Cookbook* (New York: Alfred A. Knopf, 1981), 540.

3. Ibid.

4. Irma S. Rombauer and Marion Rombauer Becker, *The Joy of Cooking* (Indianapolis: Bobbs-Merrill, 1964), 505.

5. Irma S. Rombauer and Marion Rombauer Becker, *The Joy of Cooking* (Indianapolis: Bobbs-Merrill, 1997), 772 et seq.

6. Sarah Tyson Rorer, *Mrs. Rorer's New Cook Book* (Philadelphia: Arnold and Co., 1902), 511; *Through Thick and Thin: Some Diet Suggestions from R. B. Davis Co.* (Hoboken, NJ, ca. 1920–1925), lose, 4; gain, 10.

7. Susan Marks, *Finding Betty Crocker: The Secret Life of America's First Lady of Food* (Minneapolis: University of Minneapolis Press, 2005), 168.

8. "Mrs. S. R. Dull," illustrated by Lucina Wakefield, *Southern Cooking* (New York: Grosset & Dunlap, 1941), 153.

9. "Our Story," White Lily Flour, http://www.whitelily.com/AboutUs; http://www.smuckers.com/family_company/brands.

10. "About Us," Hungry Jack, http://www.hungryjack.com/about-us.

11. The leavener was baking powder made with baking soda, sodium aluminum phosphate, and monocalcium phosphate. "Self Rising Flour," Martha White, Products, http://www.marthawhite.com/products/ProductDetail.aspx?catID=289&prodID=665.

12. "Cornbread," Martha White, Products, https://www.marthawhite.com/baking-products/cornbread; "Biscuits," Martha White, Products, https://www.marthawhite.com/baking-products/biscuit; "Muffins," Martha White, Products, https://www.marthawhite.com/baking-products/muffin.

13. "It's Cornbread Time!" Martha White, "Our Story," http://www.marthawhite.com/who-is-martha-white.

14. "Live at the Grand Ole Opry," Martha White, "Our Story," ibid.

15. "Martha White Theme Lyrics," *Metrolyrics*, Flat and Scruggs Lyrics, http://www.metrolyrics.com/martha-white-theme-lyrics-flatt-and-scruggs.html.

16. Ford: "America's Favorite Pea Picker," and "Live at the Grand Ole Opry," Martha White, "Our Story," http://www.marthawhite.com/who-is-martha-white.

17. Thanks to Randi Sunshine for pointing this out. The show began on July 6, 1974. "A Prairie Home Companion" with Chris Thile, *Prairie Home Companion*, http://prairiehome.publicradio.org/about.

18. Ibid.

19. Ibid.

20. "Our History," International Biscuit Festival, http://www.biscuitfest.com/our-history.

21. "Southern Biscuit Biscuit Baking Contest," International Biscuit Festival, http://www.biscuitfest.com/events.

22. International Biscuit Festival, http://biscuitfest.com.

23. "It's a Chemical, True Enough but Foods Are Better with It," *Food Industries* (April 1950): 81.

24. Dunkin' Brands Group, Inc., Form 10K, 2011, U.S. Securities and Exchange Commission, https://www.sec.gov/Archives/edgar/data/1357204/000119312512078208/d259421d10k.htm.

25. "Nutrition, Classic English Muffin," McDonalds.com, https://www.mcdonalds.com/us/en-us/about-our-food/nutrition-calculator.html.

26. "What Made Us Great Is Still What Makes Us Great," Kentucky Fried Chicken, https://www.kfc.com/about.

27. "Nutritional Information for your Donuts," Dunkin' Donuts, https://www.dunkindonuts.com/content/dunkindonuts/en/menu/food/bakery/donuts/donuts.html?DRP_FLAVOR=Butternut%20Donut.

28. "IHOP History," IHOP, http://www.ihop.com/about-ihop/history.

29. Melissa Kravitz, "Secrets of Magnolia Bakery: Cupcakes, Carrie Bradshaw and beyond," AM New York, May 7, 2016, http://www.amny.com/secrets-of-new-york/secrets-of-magnolia-bakery-cupcakes-carrie-bradshaw-and-beyond-1.11744934; *Sex & the City*, season 3, episode 5, "No Ifs, Ands or Butts," July 9, 2000, HBO.

30. *Cupcake Wars*, "George & Ann Lopez Charity Golf Tournament," June 13, 2010, Food Network.

31. "About Candace Nelson," Sprinkles, http://sprinkles.com/about/candace-nelson/photos.

32. Simone Beck, Louisette Bertholle, Julia Child, *Mastering the Art of French Cooking* (New York: Alfred A. Knopf, 1961), 634.

33. Gaston Lenôtre, *Lenôtre's Desserts and Pastries*, revised and adapted by Philip and Mary Hyman (Woodbury, NY: Flammarion, 1975), 100, 107, 134.

34. Tribù Golosa, http://www.tribugolosa.com.

35. Anne Volokh with Mavis Manus, *The Art of Russian Cooking* (New York: Macmillan, 1983), 470, 529.

36. Dr. Galleon, "Speculoos (Belgian Spice Cookies)," Food.com, http://www.food.com/recipe/speculoos-belgian-spice-cookies-464550.

37. Janet Mendel, "A Royal Recipe for Empanada," *My Kitchen in Spain*, October 31, 2015, http://mykitcheninspain.blogspot.com/2015/10/a-royal-recipe-for-empanada.html.

38. Glen Tancott, "Another Sweet Success," October 15, 2013, Transport World Africa, News, http://www.transportworldafrica.co.za/2013/10/15/another-sweet-success.

39. David Dolan, "South Africa's 'Spaza' Shops Suffer as Big Retail Rolls In," Business News, April 20, 2014, Reuters, http://www.reuters.com/article/2014/04/20/us-safrica-retail-spaza-idUSBREA3J06420140420.

40. "Royal Baking Powder Changes Lives," May 1, 2010, Fastmoving, http://www.fastmoving.co.za/activities/royal-baking-powder-changes-lives-211.

41. Michael Moss, *Salt Sugar Fat: How the Food Giants Hooked Us* (New York: Random House, 2013), xx.

42. Ibid., xxv.

43. *Betty Crocker's 101 Delicious Bisquick Creations as Made and Served by Well-Known Gracious Hostesses, Famous Chefs, Distinguished Epicures, and Smart Luminaries of Movieland* ([N.p.] General Mills, 1933), back cover.

44. Products, Bisquick, Betty Crocker, http://www.bettycrocker.com/products/bisquick.

45. Steve Pardo, "Happy 85th Birthday, Jiffy Mix," *Detroit News*, April 30, 2015, http://www.detroitnews.com/story/life/food/2015/04/30/jiffy-mix-biscuits-recipe/26600727.

46. "Mabel White Holmes," Michigan Women's Hall of Fame, http://www.michiganwomenshalloffame.org.

47. Betty Friedan, *The Feminine Mystique* (New York: Dell, 1983), 234.

48. *101 Delicious Bisquick Creations*, back cover.

49. Michael Pollan, *Cooked: A Natural History of Transformation* (New York: Penguin Books, 2014), 430.

50. Rachel Laudan, *Cuisine and Empire: Cooking in World History* (Berkeley: University of California Press, 2013); Rachel Laudan, "A Plea for Culinary Modernism: Why We Should Love New, Fast, Processed Food," *Gastronomica* 1, no. 1 (2001): 36–44.

51. Andrew Carnegie, *The "Gospel of Wealth," Essays and Other Writings*, edited by David Nasaw (New York: Penguin Books, 2006), 527.

52. Giuseppe Rudmani, comp., *The Royal Baker and Pastry Cook* (New York: Royal Baking Powder Co., 1878).

53. Marjorie Kinnan Rawlings, *Cross Creek Cookery* (1942; New York: Simon & Schuster, 1996), 161.

54. Lillie S. Lustig, S. Claire Sondheim, Sarah Rensel, comp. and ed., *The Southern Cook Book of Fine Old Recipes* (Reading, PA: Culinary Arts Press, 1939), 46.

55. Women of the First Congregational Church, Marysville, Ohio, *Centennial Buckeye Cookbook*, introduction and appendices by Andrew F. Smith (Columbus: Ohio State University Press, 2000), 31.

56. Lydia Maria Child, "Over the River and Through the Wood," originally published in 1844 in *Flowers for Children*, vol. 2. Child originally wrote, "Grandfather's house," but evidently "Grandmother" has more appeal.

57. Lustig, Sondheim, and Rensel, *Southern Cook Book*, 46.

58. Michael Pollan, *In Defense of Food: An Eater's Manifesto* (New York: Penguin Books, 2008), 148.

59. Joseph Campesi, "The Joy of Cooking: Slow Food and Borgmann's 'Culture of the Table,'" *Food, Culture & Society* 16, no. 3. (2013): 414–15.

60. Stephanie, aka Girl versus Dough, October 1, 2012, recipe posted on *Tablespoon*, http://www.tablespoon.com/recipes/salted-caramel-chocolate-dump-cake/5c989f6f-b488-4ca1-8665-f9b95f21c1ae.

61. Stephanie Wise, *Quick Bread Love*, http://www.girlversusdough.com/quick-bread-love.

62. Ibid.

63. "Michael Voltaggio Makes Bread in the Microwave," Jan. 18, 2010, Youtube, https://www.youtube.com/watch?v=5gAcL9qSDLE.

64. "Harvard Students Invent Cake You Can Spray from a Can," Boston.com, http://www.boston.com/food-dining/food/2014/07/18/harvard-students-invent-cake-you-can-spray-from-can/JFWbBEnsfRGFJ2W1CFbfaI/story.html.

65. Walmart, http://www.walmart.com/ip/Argo-Double-Acting-Baking-Powder-12-oz/10789500.

66. "About Us," E. Matilda Ziegler Foundation for the Blind, Inc., http://emzfoundation.com.

67. Ibid.

68. "W. M. Wright Left $60,000,000 to Kin," *New York Times*, Sept. 10, 1931, 14.

69. Ann Hagedorn Auerbach, *Wild Ride: The Rise and Tragic Fall of Calumet Farm, Inc., America's Premier Racing Dynasty* (1994; New York: Henry Holt, 1995), 57–58; Clabber Girl, 10.

70. Ibid., 10.

71. "Disastrous Fire," Vigo County Public Library, Historical Events, www.vigo.lib.in.us/subjects/genealogy/timeline/historical.

72. Darrin Dawson, "Clabber Girl to Undergo Expansion, Restoration," Oct. 1, 2006, ISU Student Media, http://www.isustudentmedia.com/article_e3999844-8382-5681-86d6-f6d27b2253cb.html.

73. "Hulman Field Airport," Vigo County Public Library, Historical Events, http://www.vigo.lib.in.us/subjects/genealogy/timeline/historical.

74. Sigur E. Whitaker, *Tony Hulman: The Man Who Saved the Indianapolis Speedway* (Jefferson, NC: McFarland, 2014).

75. Janet Ogle-Mater, "History in a 'Jiffy,'" *Chelsea Standard*, July 3, 2008, http://www.jiffyfoodservice.com/images/pdf/History_in_a__Jiffy_.pdf.

Bibliography

Abbreviations Used

BPC Morrison, A. Cressy. *The Baking Powder Controversy*. Reproduction from Harvard Law School Library. New York: American Baking Powder Association, 1904–1907. Also "The Making of Modern Law," a collection of legal archives, and BiblioLife Network.

RCW Rumford Chemical Works Archive. Rhode Island Historical Society. Providence, Rhode Island.

Archives, Government Sources, and Special Collections

Bureau of Labor Statistics, Eleventh Annual Report. Missouri State Archives. Jefferson City, MO: Tribune Printing Co., 1889.

Laws of Missouri, Fortieth General Assembly. Jefferson City, MO: Tribune Printing Co., 1899. S.B. 887.

Markle, A. R., and Gloria M. Collins. *The House of Hulman: A Century of Service, 1850–1950*. Terre Haute, IN: Clabber Girl, 1952. Private document.

Nahum (Nach) Waxman Collection of Food and Culinary Trade Cards. Division of Rare and Manuscript Collections. Cornell University Library.

NIRA Hearing on Code of Fair Practices and Competition Presented by the Baking Powder Industry, held on April 30, 1934. Washington, DC: J. L. Ward, of Ward & Paul, official reporter, 1934.

NIRA Hearing before National Compliance Council, In the matter of Standard Brands Incorporated, June 12, 1934. 2 vols. Washington, DC: J. L. Ward, of Ward & Paul, official reporter, 1934.

U.S. Congress. *Congressional Serial Set*. Washington, DC: U.S. Government Printing Office, 1901.

U.S. Congress, Senate. Committee on Agriculture and Forestry, Pure Food Legislation: "Memorial of the American Baking Powder Association in the Matter of the Bill S. 3618." 56th Cong., 1st sess., April 21, 1900.

U.S. Congress, Senate. Committee on Manufactures. "Adulteration of Food Products." 56th Congress, 1st sess., Feb. 28, 1900.

U.S. Congress, Senate. Committee on Manufactures. "Review and Topical Digest of the Evidence Taken before the Senate Committee on Manufactures between March 1899 and Feb. 1900." 56th Cong., 2nd sess., Dec. 6, 1900.

Law Suits

In the Matter of the Rendition of Wm. Ziegler to Missouri. New York, 1903.

People of the State of California v. J. A. Saint (Police Court of Los Angeles, CA, December 1903), BPC, 952–64.

People of the State of California v. A. Zimbelman (Police Court of Los Angeles, CA, December 1903), BPC, 952–64.

Rumford Chemical Works v. Hygienic Chemical Co. of New York, 215 U.S. 156 (1909). U.S. S. Ct.

Shawnee Milling Co. v. Temple, U.S. Dist. Atty., et al. (S. D. Iowa, C. D. May 10, 1910) 179 F. 117 Rptr., 520.

State of Missouri v. Whitney Layton, Missouri Supreme Court, Oct. 1900. BPC.

State of Missouri v. Whitney Layton, St. Louis Court of Criminal Correction, City of St. Louis, Missouri, Defendant's Brief. BPC.

Wabash, St. Louis & Pacific Railway Co. v. Illinois. *The Oxford Companion to the Supreme Court of the United States*, edited by Kermit L. Hall. Oxford: Oxford University Press, 2005.

Whitney Layton v. Missouri. S. Ct. No. 69, 1902.

Ziegler vs. Hoagland, et al. (NY 1888).

Books

Accum, Frederick. *A Treatise on Adulterations of Food and Culinary Poisons*. Philadelphia: Ab'm Small, 1820.

Alcott, William Andrus. *The Young House-Keeper or Thoughts on Food and Cookery*, 6th ed. Boston: Waite Peirce, and Co., 1846.

Alum in Baking Powder: The Complete Text of the "Trial Examiner's Report Upon the Facts" . . . in the Matter of Royal Baking Powder Company, Docket No. 540. New York: Royal Baking Powder Company, 1927.

Ambrose, Stephen E. *Institutions in Modern America: Innovation in Structure and Process*. Baltimore: Johns Hopkins University Press, 1967.

Anderson, Oscar E., Jr. *The Health of a Nation: Harvey W. Wiley and the Fight for Pure Food*. Chicago: University of Chicago Press, 1958.

Atkinson, Edward. *Suggestions Regarding the Cooking of Food*. Intro. Mrs. Ellen H. Richards. Washington: USDA, Government Printing Office, 1894.

Auerbach, Ann Hagedorn. *Wild Ride: The Rise and Tragic Fall of Calumet Farm, Inc., America's Premier Racing Dynasty*. New York: Henry Holt, 1995.

Barnes, Donna R., and Peter G. Rose. *Matters of Taste: Food and Drink in Seventeenth-Century Dutch Art and Life*. Albany, NY: Albany Institute of History and Art, 2002.

Beatty, Jack, ed. *Colossus: How the Corporation Changed America*. New York: Broadway Books, 2001.
Beck, Simone, Louisette Bertholle, Julia Child, *Mastering the Art of French Cooking*. New York: Alfred A. Knopf, 1961.
Beetham, Margaret. "Of Recipe Books and Reading in the Nineteenth Century: Mrs. Beeton and her Cultural Consequences." In *The Recipe Reader: Narratives, Contexts, Traditions,"* edited by Janet Floyd and Laurel Forster. Lincoln: University of Nebraska Press, 2003.
Belasco, Warren, and Roger Horowitz, eds. *Food Chains: From Farmyard to Shopping Cart*. Philadelphia: University of Pennsylvania Press, 2009.
Belasco, Warren, and Philip Scranton, eds. *Food Nations: Selling Taste in Consumer Societies*. London: Routledge, 2001.
Blaszczyk, Regina Lee, and Philip B. Scranton. *Major Problems in American Business History*. Boston: Houghton Mifflin, 2006.
Blee, Kathleen M. *Women of the Klan: Racism and Gender in the 1920s*. Berkeley: University of California Press, 1991.
Bower, Anne L. *Recipes for Reading: Community Cookbooks, Stories, Histories*. Amherst: University of Massachusetts Press, 1997.
Brewer, Priscilla J. *From Fireplace to Cookstove: Technology and the Domestic Ideal in America*. Syracuse, NY: Syracuse University Press, 2000.
Brown, Kathleen. "The Maternal Physician: Teaching American Mothers to Put the Baby in the Bathwater." In *Right Living: An Anglo-American Tradition of Self-Help Medicine and Hygiene*, edited by Charles E. Rosenberg. Baltimore: Johns Hopkins University Press, 1993.
Burnett, Frances Hodgson. *Little Lord Fauntleroy*, 2nd ed. New York, 1889.
Bryant, Keith L., Jr. and Henry C. Dethloff. *A History of American Business*, 2nd ed. Englewood Cliffs, NJ: Prentice Hall, 1990.
Cadbury, Deborah. *Chocolate Wars: The 150-Year Rivalry between the World's Greatest Chocolate Makers*. New York: Public Affairs, 2010.
Calumet Baking Powder Company. *Why White of Eggs: A Truthful Sketch of the Inception and Efficiency of the Use of White of Eggs in Baking Powder; of the Recent Attempts to Restrict Its Use by False Interpretation of the Food Laws; and a Word as to the Future*. Chicago: Calumet Baking Powder Company, 1914.
Carnegie, Andrew. *The "Gospel of Wealth," Essays, and Other Writings*, edited by David Nasaw. New York: Penguin Books, 2006.
Chandler, Alfred D., Jr. *The Visible Hand: The Managerial Revolution in American Business*. Cambridge: Belknap Press, 1977.
——, and Richard S. Tedlow. *The Coming of Managerial Capitalism: A Casebook on the History of American Economic Institutions*. Homewood, IL: Irwin, 1985.
Cogan, Thomas. *The Haven of Health*. London: Printed by Anne Griffin, for Roger Ball, 1636.
Cohen, Kim. "'True and Faithful in Everything': Recipes for Servant and Class Reform in Catherine Owen's Cookbook Novels." In *Culinary Aesthetics and Practices in Nineteenth-Century American Literature*, edited by Monika Elbert and Marie Drews. New York: Palgrave Macmillan, 2009.
Coppin, Clayton A., and Jack High. *The Politics of Purity: Harvey Washington Wiley and the Origins of Federal Food Policy*. Ann Arbor: University of Michigan Press, 1999.
Corey, Lewis. *The House of Morgan: A Social Biography of the Masters of Money*. New York: G. Howard Watt, 1930.

Cott, Nancy. *Bonds of Womanhood: "Woman's Sphere" in New England, 1780–1835*. New Haven, CT: Yale University Press, 1977.

Cowan, Ruth Schwartz. *More Work for Mother: The Ironies of Household Technology from the Open Hearth to the Microwave. New York:* Basic Books, 1983.

Cronon, William. *Changes in the Land: Indians, Colonists, and the Ecology of New England*. New York: Hill and Wang, 1983.

Cross, Gary. *An All-Consuming Century: Why Commercialism Won in Modern America*. New York: Columbia University Press, 2000.

Dahnke, Michael D., and H. Michael Dreher. *Philosophy of Science for Nursing Practice: Concepts and Application*. New York: Springer Publishing, 2011.

Dalby, Andrew. *Flavours of Byzantium*. Devon, Eng: Prospect Books, 2003.

Darrah, Juanita E. *Modern Baking Powder: An Effective, Healthful Leavening Agent*. Chicago: Commonwealth Press, 1927.

Davis, Natalie Zemon. *Women on the Margins: Three Seventeenth-Century Lives*. Cambridge: Harvard University Press, 1995.

De la Peña, *Empty Pleasures: The Story of Artificial Sweeteners from Saccharin to Splenda*. Chapel Hill: University of North Carolina Press, 2010.

Deadly Adulteration and Slow Poisoning Unmasked; or, Disease and Death in the Pot and the Bottle: in Which the Blood-Empoisoning and Life-Destroying Adulterations of Wines, Spirits, Beer, Bread, Flour, Tea, Sugar, Spices, Cheese-mongery, Pastry, Confectionary Medicine, etc. etc. etc. Are Laid Open to the Public. London: Sherwood, Gilbert, and Piper, ca. 1830.

Elliott, Clark A., and Margaret W. Rossiter. *Science at Harvard University: Historical Perspectives*. Bethlehem, PA: Lehigh University Press, 1992.

Emerson, Ralph Waldo. "History." *Self-Reliance and Other Essays*. New York: Dover Publications, 1993.

Farmer, A. N., and Janet Rankin Huntington. *Food Problems: To Illustrate the Meaning of Food Waste and What May Be Accomplished by Economy and Intelligent Substitution*. Boston: Ginn and Co., 1918.

Finley, Ruth E. *The Lady of Godey's: Sarah Josepha Hale*. Philadelphia: J. B. Lippincott, 1931.

Floyd, Janet, and Laurel Forster, eds. *The Recipe Reader: Narratives, Contexts, Traditions*. 2003. Lincoln: University of Nebraska Press, 2010.

Fox, Richard Wightman, and T. J. Jackson Lears. *The Culture of Consumption: Critical Essays in American History, 1880–1980*. New York: Pantheon Books, 1983.

Friedan, Betty. *The Feminine Mystique*. New York: Dell, 1983.

Gamber, Wendy. *The Boardinghouse in Nineteenth-Century America*. Baltimore: Johns Hopkins University Press, 2007.

Garvey, Ellen Gruber. *The Adman in the Parlor: Magazines and the Gendering of Consumer Culture, 1880s to 1910s*. New York: Oxford University Press, 1996.

Glaeser, Edward L., and Claudia Goldin, eds. *Corruption and Reform: Lessons from America's Economic History*. Chicago: University of Chicago Press, 2006.

Goldberg, David J. *Discontented America: The United States in the 1920s*. Baltimore: Johns Hopkins University Press, 1999.

Goldstein, Carolyn M. *Creating Consumers: Home Economists in Twentieth-Century America*. Chapel Hill: University of North Carolina Press, 2012.

Goodwin, Lorine Swainston. *The Pure Food, Drink, and Drug Crusaders, 1879–1914*. Jefferson, NC: McFarland, 1999.

Gordon, Maggie. *The Gilded Age on Connecticut's Gold Coast: Transforming Greenwich, Stamford, and Darien*. Charleston, SC: History Press, 2014.

Graham, Sylvester. *A Treatise on Bread and Bread-Making*. Boston: Light & Stearns, 1837.

Gray, James. *Business without Boundary: The Story of General Mills*. Minneapolis: University of Minnesota Press, 1954.

Hale, Sarah J., and Louis A. Godey, eds. *Godey's Lady's Book and Magazine* 88, Jan.–June 1874. Philadelphia: Louis A. Godey, 1874.

Hartshorn, Peter. *I Have Seen the Future: A Life of Lincoln Steffens*. Berkeley: Counterpoint, 2011.

Haynes, Roslynn. "The Alchemist in Fiction: The Master Narrative." In *The Public Image of Chemistry*, edited by Joachim Schummer, Bernadette Bensaude-Vincent, and Brigitte Van Tiggelen, 7–36. Singapore: World Scientific Publishing, 2007.

Helbich, Wolfgang, and Walter D. Kamphoefner, eds. *German-American Immigration and Ethnicity in Comparative Perspective*. Madison: University of Wisconsin, Max Kade Institute for German-American Studies, 2004.

Hess, John L., and Karen Hess. *The Taste of America*. Urbana: University of Illinois Press, 2000.

The History of Jasper County, Missouri; History of Carthage and Joplin. Des Moines, IA: Mills & Co., 1883.

Hitchcock, Edward. *Dyspepsy Forestalled and Resisted: or Lectures on Diet, Regimen, and Employment; delivered to the students of Amherst College, Spring Term, 1830*. Amherst, MA: J. S. & C. Adams, 1831.

Hoebeke, C. H. *The Road to Mass Democracy: Original Intent and the Seventeenth Amendment*. New Brunswick, NJ: Transaction, 1995.

Horsford, Eben N. *The Theory and Art of Bread-Making: A New Process without the Use of Ferment*. Cambridge, MA: Welch, Bigelow, and Co., 1861.

Howe, Daniel Walker. *Making the American Self: Jonathan Edwards to Abraham Lincoln*. Cambridge, MA: Harvard University Press, 1997.

———. *What Hath God Wrought: The Transformation of America, 1815–1848*. Oxford: Oxford University Press, 2007.

Jaffe, Bernard. *Men of Science in America: The Role of Science in the Growth of Our Country*. New York: Simon and Schuster, 1946.

Jay, Robert. *The Trade Card in Nineteenth-Century America*. Columbia: University of Missouri Press, 1987.

Johnson, Paul E. *A Shopkeeper's Millennium*. 1978. New York: Hill and Wang, 2004.

Josephson, Matthew. *The Robber Barons: The Great American Capitalists, 1861–1901*. 1934. New York: Harvest, 1962.

Kathrens, Michael C. *Great Houses of New York, 1880–1930*. New York: Acanthus Press, 2005.

Kazal, Russell A. *Becoming Old Stock: The Paradox of German-American Identity*. Princeton, NJ: Princeton University Press, 2004.

Keller, Morton. *Regulating a New Economy: Public Policy and Economic Change in America, 1900–1933*. Cambridge, MA: Harvard University Press, 1990.

Kennedy, David M., Lizabeth Cohen, and Thomas A. Bailey. *The American Pageant: A History of the Republic*. 12th ed. Vol. 1. New York: Houghton Mifflin, 2002.

Kens, Paul. *Lochner v. New York: Economic Regulation on Trial*. Lawrence: University Press of Kansas, 1998.

Kern-Foxworth, Marilyn. *Aunt Jemima, Uncle Ben, and Rastus: Blacks in Advertising, Yesterday, Today, and Tomorrow*. Westport, CT: Praeger, 1994.

Klieger, P. Christiaan. *Images of America: The Fleischmann Yeast Family*. Charleston SC: Arcadia Publishing, 2004.

Koehn, Nancy F. *Brand New: How Entrepreneurs Earned Consumers' Trust from Wedgwood to Dell*. Boston: Harvard Business School Press, 2001.

Korsmeyer, Carolyn. *Making Sense of Taste: Food & Philosophy*. Ithaca, NY: Cornell University Press, 1999.

Lamoreaux, Naomi R. *The Great Merger Movement in American Business, 1895–1904*. Cambridge, UK: Cambridge University Press, 1985.

Lang, Barbara. *The Process of Immigration in German-American Literature from 1850 to 1900: A Change in Ethnic Self-Definition*. München: Wilhelm Fink Verlag, 1988.

Laudan, Rachel. *Cuisine and Empire: Cooking in World History*. Berkeley: University of California Press, 2013.

Law, Marc T., and Gary D. Libecap, "The Determinants of Progressive Era Reform: The Pure Food and Drug Act of 1906." In *Corruption and Reform: Lessons from America's Economic History*, edited by Edward L. Glaeser and Claudia Goldin. Chicago: University of Chicago Press, 2006.

Lears, Jackson. *Fables of Abundance: A Cultural History of Advertising in America*. New York: Basic Books, 1994.

——. *No Place of Grace: Antimodernism and the Transformation of American Culture, 1880–1920*. New York: Pantheon Books, 1981.

——. *Rebirth of a Nation: The Making of Modern America, 1877–1920*. New York: Harper Perennial, 2009.

Levenstein, Harvey A. *Fear of Food: A History of Why We Worry about What We Eat*. Chicago: University of Chicago Press, 2012.

——. *Paradox of Plenty: A Social History of Eating in Modern America*. New York: Oxford University Press, 1993.

——. *A Revolution at the Table: The Transformation of the American Diet*. New York: Oxford University Press, 1988.

Levinson, Marc. *The Great A&P and the Struggle for Small Business in America*. New York: Hill and Wang, 2011.

Lewis, Sinclair. *Elmer Gantry*. Afterword by Mark Schorer. 1927. New York: Signet Classic, 1970.

Liebig, Justus von. *Researches on the Chemistry of Food and the Motion of the Juices in the Animal Body*. Lowell, MA: Daniel Bixby and Co., 1848.

Lipartito, Kenneth, and David B. Sicilia, eds. *Constructing Corporate America: History, Politics, Culture*. Oxford: Oxford University Press, 2004.

Longone, Janice (Jan) Bluestein. "Cookbooks and Manuscripts: From the Beginnings to 1860." In *The Oxford Encyclopedia of Food and Drink in America*, vol. 1, edited by Andrew F. Smith. New York: Oxford University Press, 2004.

Lyon, Peter. *Success Story: The Life and Times of S. S. McClure*. New York: Charles Scribner's Sons, 1963.

Manring, M. M. *Slave in a Box: The Strange Career of Aunt Jemima*. Charlottesville: University of Virginia Press, 1998.

Marks, Susan. *Finding Betty Crocker: The Secret Life of America's First Lady of Food*. Minneapolis: University of Minneapolis Press, 2005.

Masquerades, Tableaux, and Drills. New York: Butterick Publishing Co., 1906.

Matthews, Glenna. *"Just a Housewife": The Rise and Fall of Domesticity in America*. New York: Oxford University Press, 1987.

McCann, Alfred W. *Starving America*. New York: George H. Doran Co., 1912.

McGee, Harold. *On Food and Cooking: The Science and Lore of the Kitchen*. New York: Scribner, 2004.

McWilliams, James E. *A Revolution in Eating: How the Quest for Food Shaped America*. New York: Columbia University Press, 2005.

Mendelson, Anne. *Stand Facing the Stove: The Story of the Women Who Gave America* The Joy of Cooking. New York: Henry Holt, 1996.

Mendelson, Simon. *Baking Powders: Including Chemical Leavening Agents, Their Development, Chemistry, and Valuation*. New York: Chemical Publishing Co., 1939.

Mintz, Sidney. *Tasting Food, Tasting Freedom: Excursions into Eating, Culture, and the Past*. Boston: Beacon Press, 1996.

Moltmann, Gunther, ed. *Germans to America: 300 Years of Immigration, 1683 to 1983*. Stuttgart: Institute for Foreign Cultural Relations, in cooperation with Inter Nationes, Bonn-Bad Godesberg, 1982.

Monkkonen, Eric H. *America Becomes Urban: The Development of U.S. Cities & Towns, 1780–1980*. Berkeley: University of California Press, 1988.

Moore, Leonard J. *Citizen Klansmen: The Ku Klux Klan in Indiana, 1921–1928*. Chapel Hill: University of North Carolina Press, 1997.

Morrison, A. Cressy. *Man in a Chemical World: The Service of Chemical Industry*. New York: Charles Scribner's Sons, 1937.

Moss, Michael. *Salt Sugar Fat: How the Food Giants Hooked Us*. New York: Random House, 2013.

Mullins, Paul R. *Glazed America: A History of the Doughnut*. Gainesville: University Press of Florida, 2008.

Nestle, Marion. *Food Politics: How the Food Industry Influences Nutrition and Health*. Berkeley: University of California Press, 2002.

Nissenbaum, Stephen. *Sex, Diet, and Debility in Jacksonian America: Sylvester Graham and Health Reform*. Chicago: Dorsey Press, 1980.

Norris, James D. *Advertising and the Transformation of American Society, 1865–1920*. New York: Greenwood Press, 1990.

Okun, Mitchell. *Fair Play in the Marketplace: The First Battle for Pure Food and Drugs*. DeKalb: Northern Illinois University Press, 1986.

Oliver, Sandra L. *Saltwater Foodways: New Englanders and Their Food, at Sea and Ashore, in the Nineteenth Century*. Mystic, CT: Mystic Seaport Museum, 1995.

Olsen, John C. *Pure Foods: Their Adulteration, Nutritive Value, and Cost*. Boston: Ginn and Co., 1911.

Panschar, William G. *Baking in America: Economic Development*. Vol. 1. Evanston, IL: Northwestern University Press, 1956.

Patterson, Martha H. *Beyond the Gibson Girl: Reimagining the American New Woman, 1895–1915*. Urbana: University of Illinois Press, 2005.

Peck, John Licinius Everett, Otto Hillock Montzheimer, and William J. Miller. *Past and Present of O'Brien and Osceola Counties, Iowa*. Vol. 1. Indianapolis: B. F. Bowen & Co., 1914.

Pendergrast, Mark. *For God, Country, and Coca-Cola: The Definitive History of the Great American Soft Drink and the Company That Makes It*. New York: Basic Books, 2000.

Piott, Steven L. *Holy Joe: Joseph W. Folk and the Missouri Idea*. Columbia: University of Missouri Press, 1997.

Pollan, Michael. *Cooked: A Natural History of Transformation*. New York: Penguin Books, 2014.

———. *In Defense of Food: An Eater's Manifesto*. New York: Penguin Books, 2008.

———. *Omnivore's Dilemma: A Natural History of Four Meals*. New York: Penguin, 2006.

Porter, Glenn. *The Rise of Big Business, 1860–1920*, 2nd ed. 1973. Arlington Heights, IL: Harlan Davidson, 1992.

Porter, Theodore M. *Trust in Numbers: The Pursuit of Objectivity in Science and Public Life*. Princeton, NJ: Princeton University Press, 1995.

Presbrey, Frank. *The History and Development of Advertising*. Garden City, NY: Doubleday, Doran & Co., 1929.

Regier, Cornelius C. *The Era of the Muckrakers*. Chapel Hill: University of North Carolina Press, 1932.

Richards, Ellen H. *The Chemistry of Cooking and Cleaning: A Manual for Housekeepers*. Boston, Estes & Lauriat, 1882.

———. *Euthenics: The Science of Controllable Environment*. 1910. New York: Arno Press, 1977.

———. *Food Materials and Their Adulterations*, 3rd ed. Boston: Whitcomb & Barrows, 1911.

Romines, Ann. "Reading the Cakes: *Delta Wedding* and the Texts of Southern Women's Culture." *Mississippi Quarterly* 50, no. 4 (1997): 601–616.

Root, Waverly, and Richard de Rochemont. *Eating in America*. Hopewell, NJ: Ecco Press, 1995.

Rosenberg, Charles E. *No Other Gods: On Science and American Social Thought*. 1976. Baltimore: Johns Hopkins University Press, 1997.

———. *Right Living: An Anglo-American Tradition of Self-Help Medicine and Hygiene*. Charles E. Rosenberg, ed. Baltimore: Johns Hopkins University Press, 1993.

Rossiter, Margaret W. *The Emergence of Agricultural Science: Justus Liebig and the Americans, 1840–1880*. New Haven, CT: Yale University Press, 1975.

———. *Science at Harvard*. Bethlehem, PA: Lehigh University Press, 1992.

Rossum, Ralph A. *Federalism, the Supreme Court, and the Seventeenth Amendment: The Irony of Constitutional Democracy*. Lanham, MD: Lexington Books, 2001.

Rubel, William. *Bread: A Global History*. Edible Series. Andrew F. Smith, ed. London: Reaktion Books, 2011.

Salmon, Lucy Maynard. *History and the Texture of Modern Life*. Nicholas Adams and Bonnie G. Smith, eds. Philadelphia: University of Pennsylvania Press, 2001.

Sarton, George. *The History of Science and the New Humanism*. New York: George Braziller, 1956.

———. *The History of Science and the Problems of To-Day*. Elihu Root Lectures of Carnegie Institution of Washington on the Influence of Science and Research on Current Thought. Washington: W. F. Roberts Co., 1936.

Schlosser, Eric. *Fast Food Nation: The Dark Side of the All-American Meal*. New York: Perennial, 2002.

Scranton, Philip. *Endless Novelty: Specialty Production and American Industrialization, 1865–1925*. Princeton, NJ: Princeton University Press, 1997.

Shapin, Steven. *Never Pure: Historical Studies of Science as if It Was Produced by People with Bodies, Situated in Time, Space, Culture, and Society, and Struggling for Credibility and Authority*. Baltimore: Johns Hopkins University Press, 2010.

Shapiro, Herbert, ed. *The Muckrakers and American Society*. Lexington, MA: D. C. Heath and Co., 1968.

Shapiro, Laura. *Perfection Salad: Women and Cooking at the Turn of the Century*. New York: Farrar, Straus & Giroux, 1986.

Shrock, Joel. *The Gilded Age*. Westport, CT: Greenwood Press, 2004.

Sklar, Kathryn Kish. *Catharine Beecher: A Study in American Domesticity*. New York: W. W. Norton & Co., 1976.

Slater, Charles C., comp. *Baking in America: Market Organization and Competition*. Evanston, IL: Northwestern University Press, 1956.

Smith, Andrew F., ed. *The Oxford Encyclopedia of Food and Drink in America*. New York: Oxford University Press, 2004.

Smith, Ernest Ellsworth. *Aluminum Compounds in Food*. New York: Paul B. Hoeber, 1928.

Steffens, Lincoln. *Autobiography*. New York: Harcourt, Brace and Co., 1931.

———. *The Letters of Lincoln Steffens*. Vol. 1: *1889–1919*. New York: Harcourt, Brace and Co., 1938.

———. *The Struggle for Self-Government: Being an Attempt to Trace American Political Corruption to Its Sources in Six States of the United States with a Dedication to the Czar*. New York: McClure, Phillips & Co., 1906.

Stephens, Carl, ed. *The Alumni Record of the University of Illinois*. Dixon, IL: Rogers Printing Co., 1921.

Stowe, Harriet Beecher. *The Minister's Wooing*. New York: Penguin, 1859.

Strasser, Susan. *Never Done: A History of American Housework*. New York: Pantheon Books, 1982.

———, Charles McGovern, and Matthias Judt, eds. *Getting and Spending: European and American Consumer Societies in the Twentieth Century*. Washington, DC: German Historical Institute, Cambridge University Press, 1998.

Stuart, Benj. F. *History of the Wabash and Valley*. [Logansport, IN]: Longwell-Cummings Co., 1924.

Swedberg, Richard. *Entrepreneurship: The Social Science View*. New York: Oxford University Press, 2000.

Tedlow, Richard. *Giants of Enterprise: Seven Business Innovators and the Empires They Built*. New York: HarperCollins Publishers, 2001.

———. *Keeping the Corporate Image: Public Relations and Business, 1900–1950, Industrial Development, and the Social Fabric.* Vol. 3. Greenwich, CT: JAI Press, 1979.

———. *New and Improved: The Story of Mass Marketing in America*. Boston: Harvard Business School Press, 1996.

Terre Haute of Today. Terre Haute, IN: Moore & Langen, 1895.

Theophano, Janet. *Eat My Words: Reading Women's Lives through the Cookbooks They Wrote*. New York: Palgrave, 2002.

The Truth about Baking Powder. Compiled and distributed by Calumet Baking Powder Co. Chicago, 1928.

Usselman, Steven W. *Regulating Railroad Innovation: Business, Technology, and Politics in America, 1840–1920*. Cambridge, UK: Cambridge University Press, 2002.

The Wabash Valley Remembers: A Chronicle, 1787–1938. Terre Haute, IN: Northwest Territory Celebration Committee, 1938.

Wallop, Harry. "Bread Rules Abandoned after 750 Years." *Telegraph*. Sept. 25, 2008.
Weaver, William Woys. *The Christmas Cook: Three Centuries of American Yuletide Sweets*. New York: HarperPerennial, 1990.
Wetmore, Claude. *The Battle against Bribery: Being the Only Complete Narrative of Joseph W. Folk's Warfare on Boodlers*. St. Louis: Pan-American Press, 1904.
Whitaker, Sigur E. *Tony Hulman: The Man Who Saved the Indianapolis Speedway*. Jefferson, NC: McFarland, 2014.
Wiley, Harvey Washington. *1001 Tests of Foods, Beverages, and Toilet Accessories, Good and Otherwise*. New York: Hearst's International Library, 1914.
———. *An Autobiography*. Indianapolis: Bobbs-Merrill, 1930.
———. *The History of a Crime against the Food Law: The Amazing Story of the National Food and Drugs Law Intended to Protect the Health of the People Perverted to Protect Adulteration of Foods and Drugs*. Washington, DC: Harvey W. Wiley, MD, 1929.
Wilson, Harold S. *McClure's Magazine and the Muckrakers*. Princeton, NJ: Princeton University Press, 1970.
Winton, Andrew L. *A Course in Food Analysis*. New York: John Wiley & Sons, 1917.
Wood, Donna. *Strategic Uses of Public Policy*. Marshfield, MA: Pitman Publishing, 1986.
Yeager, Mary A., ed. *Competition and Regulation: The Development of Oligopoly in the Meat Packing Industry*. Greenwich, CT: JAI Press, 1981.
———. *Women in Business,* Vol. 1. Northampton, MA: Edward Elgar Publishing, 1999.
Yeatman, Ted P. *Frank and Jesse James*. Nashville: Cumberland House Publishing, 2000.
Yellin, Jean Fagan. *Women & Sisters: The Antislavery Feminists in American Culture*. New Haven, CT: Yale University Press, 1989.
Young, James Harvey. *Pure Food: Securing the Federal Food and Drugs Act of 1906*. Princeton, NJ: Princeton University Press, 1989.
———. *The Toadstool Millionaires: A Social History of Patent Medicines in America before Federal Regulation*. Princeton, NJ: Princeton University Press, 1961.
Zunz, Olivier. *Making America Corporate, 1870–1920*. Chicago: University of Chicago Press, 1990.

Cookbooks

Abel, Mary Hinman. *Practical Sanitary and Economic Cooking Adapted to Persons of Moderate and Small Means*. New York: American Public Health Association, 1890.
Accademia Italiana in Cucina (Italian Academy of Cuisine). *La Cucina: The Regional Cooking of Italy*. New York: Rizzoli, 2009.
Anon. *The Cook Not Mad*. Watertown, NY: Knowlton & Rice, 1831.
Beard, James. *The New James Beard Cookbook*. New York: Alfred A. Knopf, 1981.
Beecher, Catharine. *Miss Beecher's Domestic Receipt-Book*. 1841. Mineola, NY: Dover, 2001.
Beeton, Isabella. *Mrs. Beeton's Book of Household Management*. London, 1861.
Betty Crocker's 101 Delicious Bisquick Creations as Made and Served by Well-Known Gracious Hostesses, Famous Chefs, Distinguished Epicures, and Smart Luminaries of Movieland. [N.p.] General Mills, 1933.
Bullock, Helen Duprey, and William Parks, *The Williamsburg Art of Cookery; or, Accomplish'd Gentlewoman's Companion: Being a Collection of Upwards of Five Hundred of the Most Ancient &*

Approv'd Recipes in Virginia Cookery . . . And Also a Table of Favorite Williamsburg Garden Herbs. Colonial Williamsburg, VA, 1942.

Calumet Cook Book. Chicago: Calumet Baking Powder Co., ca. 1923.

Child, Lydia Maria. *The American Frugal Housewife.* 1829. Mineola, NY: Dover Publications, 1999.

Clabber Girl Baking Book. Terre Haute, IN: Hulman & Co., ca. 1932.

Clabber Girl: The Healthy Baking Powder. Terre Haute, IN: Hulman & Co., 1931.

The Cook's Book: Recipes Prepared for KC Baking Powder. Chicago: Jaques Mfg. Co., ca. 1939.

Davidis, Henriette. *Pickled Herring and Pumpkin Pie: A Nineteenth-Century Cookbook for German Immigrants to America.* Intro. Louis A. Pitschmann. Madison, WI: Max Kade Institute for German-American Studies, 2003.

De Voe, Thomas Farrington. *The Market Assistant, Containing a Brief Description of Every Article of Human Food Sold in the Public Markets of the Cities of New York, Boston, Philadelphia, and Brooklyn; Including the Various Domestic and Wild Animals, Poultry, Game Fish, Vegetables, Fruits, &c, &c. with many Curious Incidents and Anecdotes.* New York: Hurd and Houghton, 1867.

Dr. Price. *Table and Kitchen: A Compilation of Approved Cooking Receipts.* Chicago, 1908.

Dull, Mrs. S. R. Illustrated by Lucina Wakefield. *Southern Cooking.* New York: Grosset & Dunlap, 1941.

Estes, Rufus. *Things to Eat, as Suggested by Rufus; A Collection of Practical Recipes for Preparing Meats, Game, Fowl, Fish, Puddings, Pastries, etc.* Chicago: The Author [c1911].

Farmer, Fannie Merritt. *The Original 1896 Boston Cooking-School Cook Book.* 1896. Boston: Dover Publications, 1997.

Fisher, Mrs. Abby. *What Mrs. Fisher Knows about Old Southern Cooking.* Facsimile, with historical notes by Karen Hess. 1881. Bedford, MA: Applewood Books, 1995.

Fox, Minnie C., comp. *The Blue Grass Cook Book.* New introduction by Toni Tipton-Martin. 1904. Lexington: University Press of Kentucky, 2005.

Gillespie, Elizabeth Duane, comp. *National Cookery Book, compiled from original receipts, for the Women's Centennial Committees of the International Exposition.* Philadelphia: Women's Centennial Executive Committee. Intro. Andrew F. Smith. 1876. Bedford, MA: Applewood Books, 2010.

Gillette, Mrs. F. L. *White House Cook Book.* Chicago: R. S. Peale & Co., 1887.

Glasse, Hannah. *The Art of Cookery Made Plain and Easy.* 1747. Bedford, MA: Applewood Books, 1997.

Gone with the Wind Cook Book: Famous "Southern Cooking" Recipes. New York: Anthony P. Cima, 1940.

Hale, Sarah Josepha. *Early American Cookery: The "Good Housekeeper."* 1841. Mineola, NY: Dover Publications, 1996.

Hall, Elizabeth M, comp. *Practical American Cookery and Domestic Economy.* New York: C. M. Saxton, Barker & Co., 1860.

Here Are the Cakes America Loves: Royal Cakes Made with Royal Cream of Tartar Baking Powder. New York: Standard Brands Inc., 1950.

How to Take a Trick a Day with Bisquick, as told to Betty Crocker. [N.p.] General Mills, 1935.

Hughes, Phyllis, ed. *Pueblo Indian Cookbook.* Santa Fe: University of New Mexico Press, 1982.

Hutchinson, Ruth. *The New Pennsylvania Dutch Cook Book.* New York: Harper & Row, 1985.

Jennie June's American Cookery Book. New York: American News Co., 1870.

Kander, Mrs. Simon, compiler. *The "Settlement" Cook Book: Containing Many Recipes Used in Settlement Cooking Classes, the Milwaukee Public School Cooking Centers, and Gathered from Various Other Reliable Sources*. Milwaukee, 1901.

Kellogg, Ella Eaton. *Science in the Kitchen. A Scientific Treatise on Food Substances and Their Dietetic Properties, Together with a Practical Explanation of the Principles of Healthful Cookery, and a Large Number of Original, Palatable, and Wholesome Recipes*. Chicago: Modern Medicine Publishing Co., 1893.

Kimball, Marie. *The Martha Washington Cook Book*. New York: Coward-McCann, 1940.

Lenôtre, Gaston. *Lenôtre's Desserts and Pastries*. Revised and adapted by Philip and Mary Hyman. Woodbury, NY: Flammarion, 1975.

Leslie, Eliza. *Miss Leslie's New Cookery Book*. Philadelphia: T. B. Peterson and Brothers, 1857.

———. *Seventy-Five Receipts for Pastry, Cakes, and Sweetmeats*, 3rd ed. Boston: Munroe and Francis, 1828.

Lustig, Lillie S., S. Claire Sondheim, and Sarah Rensel, comp. and ed. *The Southern Cook Book of Fine Old Recipes*. Reading, PA: Culinary Arts Press, 1939.

Martha Washington's Booke of Cookery and Booke of Sweetmeats. Transcribed by Karen Hess with historical notes and copious annotations. New York: Columbia University Press, 1981.

Mérigot, Mme. *La Cuisiniere Républicaine, Qui enseigne la manière simple d'accommoder les Pommes de terre*. Paris: Chez Mérigot jeune, libraire, L'an III de la République [1794 or 1795]. In Beatrice Fink, ed., *Les Liaisons Savoureuses: Réflexions et pratiques culinaires au XVIIIe Siècle*. Saint-Étienne, France: Publications de l'Université de Saint-Étienne, 1995.

The Metropolitan Life Cook Book. [n.p.] Metropolitan Life Insurance Co., 1922.

Morris, William. *Professor Morris' Azumea, the Premium Baking Powder*. Philadelphia: E.W.P. Taunton, 1867.

Murrey, Thomas Jefferson. *Breakfast Dainties*. Bedford, MA: Applewood Books, n.d. [1885].

Murphy, Claudia Quigley. *A Collation of Cakes Yesterday and Today, in Which Is Included a True and Accurate Notation of Early English and Colonial America Receipts, Showing the Beginning and Progress of the Gentle Art of Cake Making to the Present Time*. New York, 1923.

Nightingale, Florence. "Taking Food." In *Directions for Cooking by Troops in Camp and Hospital*. Richmond, VA: J. W. Randolph, 1861.

Pinedo, Encarnación. *Encarnación's Kitchen: Mexican Recipes from Nineteenth-Century California*, edited and translated by Dan Strehl. Berkeley: University of California Press, 1898.

Platina. *De honesta voluptate et valetudine* (On Right Pleasure and Good Health). In *The Art of Cooking: The First Modern Cookery Book*, edited by Luigi Ballerini, translated by Jeremy Parzen. Berkeley: University of California Press, 2005.

Purefoy, Eva Brunson. *The Purefoy Hotel Cook Book*. Talladega, AL, 1938.

Randolph, Mary. *The Virginia Housewife; or, Methodical Cook*. 1831. New York: Dover, 1993.

Recipes That Are Business Builders for Restaurants, Hotels, and Institutions. Standard Brands of California, n.d.

Reliable Recipes. Chicago: Calumet Baking Powder Co., 1909.

Rombauer, Irma S. *The Joy of Cooking*. Foreword by Edgar Rombauer. 1931. New York: Scribner, facsimile 1998.

———. Illustrations by Marion Rombauer Becker. *The Joy of Cooking*. Indianapolis: Bobbs-Merrill, 1943.

———, and Marion Rombauer Becker. *The Joy of Cooking*. Indianapolis: Bobbs-Merrill, 1953.

——, and Marion Rombauer Becker. *The Joy of Cooking*. Indianapolis: Bobbs-Merrill, 1964.

——, and Marion Rombauer Becker. *The Joy of Cooking*. Indianapolis: Bobbs-Merrill, 1975.

——, Marion Rombauer Becker, and Ethan Becker. *The Joy of Cooking*. New York: Scribner, 1997.

Rorer, Sarah Tyson. *Mrs. Rorer's New Cook Book: A Manual of Housekeeping*. Philadelphia: Arnold and Co., 1902.

Rudmani, G., comp. *The Royal Baker and Pastry Cook*. New York: Royal Baking Powder Co., 1878.

The Royal Baker and Pastry Cook. Royal Baking Powder, 1902.

Royal Cakes. New York: Standard Brands Inc., 1950.

The Royal Guide to Meal Planning. Compiled by the Educational Department of Standard Brands Incorporated. New York: Standard Brands, 1929.

Rundell, Maria Eliza Ketelby. *A New System of Domestic Cookery*. Philadelphia: Benjamin C. Buzby, 1807.

Ryzon Baking Book: A Practical Manual for the Preparation of Food Requiring Baking Powder. New York: General Chemical Company Food Department, 1918.

Shuman, Carrie V., compiler. *Favorite Dishes: A Columbian Autograph Souvenir Cookery Book*. Introductions by Reid Badger and Bruce Kraig.1893. Urbana: University of Illinois Press, 2001.

Simmons, Amelia. *American Cookery.* 2nd ed. Hartford, CT: Printed for Simeon Butler, Northampton, 1798.

——. *The First American Cookbook: A Facsimile of "American Cookery."* 1st ed., 1796. New York: Dover Publications, 1958.

The Swedish-American Cookbook: A Charming Collection of Traditional Recipes Presented in Both Swedish and English. 1882. New York: Skyhorse Publishing, 2012.

Swell, Barbara. *The 1st American Cookie Lady*. Asheville, NC: Native Ground Music, 2005.

Thompson, Ruth Plumly. *Billy in Bunbury*. Royal Baking Powder, 1925.

——. *The Comical Cruises of Captain Cooky*. Royal Baking Powder, 1926.

——. *The Little Gingerbread Man*. Royal Baking Powder, 1923.

——. *Prince of the Gelatin Isles*. Royal Baking Powder, 1926.

Through Thick and Thin: Some Diet Suggestions from R. B. Davis Co. Hoboken, NJ, ca. 1920–1925.

Volokh, Anne, with Mavis Manus. *The Art of Russian Cooking*. New York: Macmillan, 1983.

Wallace, Lily Haxworth. *The Rumford Complete Cook Book*. 1908. Providence, RI: Rumford Chemical Works, 2000.

Ward, Artemas. *The Grocer's Encyclopedia*. New York, 1911.

Women of the First Congregational Church, Marysville, Ohio. *Centennial Buckeye Cook Book*. Andrew F. Smith, intro. and appendices. 1876. Columbus: Ohio State University Press, 2000.

Plus proprietary cookbooks by Calumet, Clabber Girl, Royal, and Rumford.

Journal Articles, Treatises, and Newspapers

"Baking Powder." *Bakers Weekly*. May 27, 1933.

"Baking Powder and Boodle." *Spice Mill*. October 1903. In *The Baking Powder Controversy*, by A. Cressy Morrison, vol. 2. New York, 1904–1907.

"Baking Powders in the United States," *New York Tribune*. In *The Analyst*, edited by G. W. Wigner and J. Muter, Royal Society of Chemistry. London: Baillière, Tindall, and Cox, 1881.

Campesi, Joseph. "The Joy of Cooking: Slow Food and Borgmann's 'Culture of the Table.'" *Food, Culture & Society* 16, no. 3. (2013): 414–15.

Carter, Harvey L. "Rural Indiana in Transition, 1850–1860." *Agricultural History* 20, no. 2 (1946): 107–108.

Chandler, Alfred D., Jr. "The Enduring Logic of Industrial Success." *Harvard Business Review*. March–April 1990.

"David Will Leave Business School." *Harvard Crimson*. June 2, 1927.

"Development and Use of Baking Powder and Baking Chemicals." U.S. Department of Agriculture, Circular No. 138. Washington, DC, Nov. 1930.

Foster, Rev. Addison P. "The Dakota Indians." *American Missionary* 38, no. 6 (1884).

Gabriel, Joseph M. "Restricting the Sale of 'Deadly Poisons': Pharmacists, Drug Regulation, and Narratives of Suffering in the Gilded Age." *Journal of the Gilded Age and the Progressive Era* 9, no. 3 (2010): 316.

"The Great American Lobby: The Typical Example of Missouri." *Leslie's Popular Monthly 56 (*August 1903): 382–93.

"History of the Federal Judiciary: The Debs Case: Labor, Capital, and the Federal Courts of the 1890s, Biographies, American Railway Union." *Federal Judicial Center*.

"Horsford's Bread Preparation." *Scientific American* 22, no. 1 (1870).

Jones, Paul R. "Justus Von Liebig, Eben Horsford, and the Development of the Baking Powder Industry." *Ambix* 40, part 2 (July 1993).

Krondl, Michael. "The War That Made Donuts a Secret Weapon." June 4, 2014. Zesterdaily .com.

Laudan, Rachel. "A Plea for Culinary Modernism: Why We Should Love New, Fast, Processed Food." *Gastronomica* 1, no. 1 (2001): 36–44.

Lonier, Terri. "Alchemy in Eden: Entrepreneurialism, Branding, and Food Marketing in the United States, 1880–1920." *Enterprise & Society* 11, no. 4 (2010): 695–708.

Lukas, Paul. "Jiffy's Secret Recipe Chelsea Milling has beaten its competition—the Pillsbury Dough Boy and Betty Crocker never laid a whisk on 'em—and plans to stay on top with two parts aw-shucks family business and one part professional management." CNN Money, Dec. 1, 2001.

"Martha Washington Tea Party." *Cambridge Chronicle* 31, no. 9 (Feb. 26, 1876).

May, Vivian. "Cross Purposes," *Southern Changes* 16, no. 2 (1994): 13–17.

McWilliams, Mark. "Good Women Bake Good Biscuits: Cookery and Identity in Ante-bellum American Fiction." *Food, Culture & Society* 10, no. 3 (2007): 388–406.

Miller, Harry. "Potash from Wood Ashes: Frontier Technology in Canada and the United States." *Technology and Culture* 21, no. 2 (1980): 187–208.

Morrison, A. Cressy. Address—"The Baking Powder Controversy." *Journal of Proceedings*, Eighth Annual Convention and International Pure Food Congress of the National Association of State Dairy and Food Departments. St. Louis, Missouri. Sept. 29, 1904.

Mott, Henry A., Jr., PhD, EM. "The Deleterious Use of Alum in Bread and Baking Powders—Alum Being Substituted for Cream of Tartar." *Scientific American*. Nov. 16, 1878.

Ogle-Mater, Janet. "History in a 'Jiffy.'" *Chelsea Standard*. July 3, 2008.

"Restricting the Sale of 'Deadly Poisons': Pharmacists, Drug Regulation, and Narratives of Suffering in the Gilded Age." *Journal of the Gilded Age and the Progressive Era* 9, no. 30 (2010): 313–36.

Rezneck, Samuel. "The European Education of an American Chemist and Its Influence in 19th-Century America: Eben Norton Horsford." *Technology and Culture* 11, no. 3 (1970): 366–88.
"Royal Baking Powder Had Record Sale in 1921." *American Food Journal* 17 (Jan. 1922): 33.
"The Royal Baking Powder Company versus the Montana State Board of Health." *American Food Journal* (June 1920): 13–14.
Schmidt, Stephen. "Cakes." In *The Oxford Encyclopedia of Food and Drink in America*, edited by Andrew F. Smith. New York: Oxford University Press, 2004.
Schling-Brodersen, Uschi. "Liebig's Role." *Ambix* 39, part 1 (March 1992): 21–31.
Sewell, Jessica. "Tea and Suffrage." *Food, Culture & Society* 2, no. 4 (2008): 487–507.
"'Trick' Selling Condemned by Federal Trade Commission." *Sales Management*, July 6, 1929, 20. Chamber of Commerce statement to members, June 21, 1929.
Warner, D. J. "Ira Remsen, Saccharin, and the Linear Model," *Ambix* 55, no. 1 (2008): 50–61.
Wija, Tantri. "Out of the Frying Pan." *Free New Mexican*. August 16, 2006.
"Won't Standardize Baking Powder Cans." *American Food Journal* 17 (Oct. 1922): 46.
Wooden, Warren. "Calcium Carbonate in Gastric Ulcer." *Journal of the American Medical Association* 89, no. 17 (1927): 1445.

Papers

Albala, Ken. "Food History through the Ages." Paper presented at Tasting Histories conference, UC Davis, Feb. 28, 2009.
Shapin, Steven. "The Medical Making of Modernity: Knowing about Our Food, Our Bodies, and Ourselves over the Past 2,000 Years." Paper presented at the New York Academy of Sciences, Feb. 2, 2011.
Smith, Andrew F. "Advertising and Promotional Cookbooklets in the Nineteenth Century." Paper presented at the Cookbook Conference, New York, Feb. 9, 2012.

Index

Page numbers in *italics* indicate photographs

LINDA CIVITELLO teaches food history in southern California. She is the author of *Cuisine and Culture: A History of Food and People*, winner of the Gourmand Award for Best Food History Book in the World in English (U.S.).

HEARTLAND FOODWAYS

From the Jewish Heartland: Two Centuries of Midwest Foodways
Ellen F. Steinberg and Jack H. Prost
Farmers' Markets of the Heartland *Janine MacLachlan*
A Perfect Pint's Beer Guide to the Heartland *Michael Agnew*
Midwest Maize: How Corn Shaped the U.S. Heartland *Cynthia Clampitt*
Local Vino: The Winery Boom in the Heartland *James R. Pennell*
Baking Powder Wars: The Cutthroat Food Fight That Revolutionized Cooking
Linda Civitello

The University of Illinois Press
is a founding member of the
Association of American University Presses.

Composed in 10.25/13 Marat Pro
with Trade Gothic display
by Kirsten Dennison
at the University of Illinois Press
Manufactured by Cushing-Malloy, Inc.

University of Illinois Press
1325 South Oak Street
Champaign, IL 61820-6903
www.press.uillinois.edu